Canadian Institute
of Resources Law

Institut canadien du
droit des ressources

ENERGY CONSERVATION LEGISLATION FOR BUILDING DESIGN AND CONSTRUCTION

Adrian J. Bradbrook
Chair of Natural Resources Law
Faculty of Law
The University of Calgary

October 1992

Faculty of Law. The University of Calgary, Calgary, Alberta. T2N 1N4 Tel. (403) 220-3200

Canadian Cataloguing in Publication Data

Bradbrook, Adrian J.
Energy conservation legislation for building design and construction

Includes bibliographical references
ISBN 0-919269-36-2

1. Energy conservation – Law and legislation. I. Canadian Institute of Resources Law. II. Title.
K3981.B72 1992 346.04'67916 C92-091769-0

Printed in Canada

CANADIAN INSTITUTE OF RESOURCES LAW

The Canadian Institute of Resources Law was incorporated in 1979 with a mandate to examine the legal aspects of both renewable and nonrenewable resources. Its work falls into three interrelated areas: research, education, and publication.

The Institute has engaged in a wide variety of research projects, including studies on oil and gas, mining, forestry, water electricity, the environment, aboriginal rights, surface rights, and the trade of Canada's natural resources.

The education function of the Institute is pursued by sponsoring conferences and short courses on particular topical aspects of resources law, and through teaching in the Faculty of Law at The University of Calgary.

The major publication of the Institute is its ongoing looseleaf service, the *Canada Energy Law Service*, published in association with Thomson Professional Publishing Canada. The results of other Institute research are published as books and discussion papers. Manuscripts submitted by outside authors are considered. The Institute publishes a quarterly newsletter, *Resources*.

The Institute is supported by the Alberta Law Foundation, the Government of Canada, and the private sector. The members of the Board of Directors are appointed by the Faculty of Law at The University of Calgary, the President of The University of Calgary, the Benchers of the Law Society of Alberta, the President of the Canadian Petroleum Law Foundation, and the Dean of Law at The University of Alberta. Additional members of the Board are elected by the appointed Directors.

All enquiries should be addressed to:

The Executive Director
Canadian Institute of Resources Law
Room 430, BioSciences Building
Faculty of Law
The University of Calgary
Calgary, Alberta, Canada
T2N 1N4

Telephone: (403) 220-3200
Facsimile: (403) 282-6182
Telex: 03-821545

INSTITUT CANADIEN DU DROIT DES RESSOURCES

L'institut canadien du droit des ressources a été constitué en 1979 et a reçu pour mission d'étudier les aspects juridiques des ressources renouvelables et non renouvelables. Son travail porte sur trois domaines étroitement reliés entre eux, soit la recherche, l'enseignement et les publications.

L'institut a entrepris une vaste gamme de projets de recherche, notamment des études portant sur le pétrole et le gaz, l'exploitation des mines, l'exploitation forestière, les eaux, l'électricité, l'environnement, les droits des autochtones, les droits de surface et le commerce des ressources naturelles du Canada.

L'institut remplit ses fonctions éducatives en commanditant des conférences et des cours de courte durée sur des sujets d'actualité particuliers en droit des ressources et par le truchement de l'enseignement à la Faculté de droit de l'Université de Calgary.

La plus importante publication de l'institut est son service de publication continue à feuilles mobiles intitulé le *Canada Energy Law Service*, publié conjointement avec Thomson Professional Publishing Canada. L'institut publie également les résultats d'autres recherches sous forme de livres et de documents d'étude. Les manuscrits soumis par des auteurs de l'extérieur sont également considérés. L'institut publié un bulletin trimestriel intitulé *Resources*.

L'institut reçoit des subventions de la Alberta Law Foundation, du gouvernement du Canada et du secteur privé. Les membres du conseil d'administration sont nommés par la Faculté de droit de l'Université de Calgary, le recteur de l'Université de Calgary, les conseillers de la Law Society of Alberta, le président de la Canadian Petroleum Law Foundation et le doyen de la Faculté de droit de l'Université d'Alberta. D'autres membres sont élus par les membres du conseil nommés.

Toute demande de renseignement doit être adressée au:

Directeur exécutif
Institute canadien du droit des ressources
Room 430, BioSciences Building
Faculty of Law
The University of Calgary
Calgary, Alberta, Canada
T2N 1N4

Téléphone: (403) 220-3200
Facsimilé: (403) 282-6182
Télex: 03-821545

ACKNOWLEDGEMENTS

This study was undertaken during the period in which the author was the holder of the Visiting Chair in Natural Resources Law at the Faculty of Law, University of Calgary. The author would like to thank Professor Owen Saunders, the Executive Director of the Canadian Institute of Resources Law, and Professor Constance D. Hunt, the Dean of the Faculty of Law at that time, for providing the opportunity and the appropriate conditions for the study to take place.

This study was edited by Evangeline Case, Publications Officer. The author wishes to express his appreciation for her attention to detail and for her helpful suggestions throughout the publication process. The author is also grateful to Sue Parsons, who prepared the typeset material.

TABLE OF CONTENTS

FOREWORD

This book reflects research carried out by Professor Bradbrook in the course of his tenure as the 1991 incumbent of the Chair of Natural Resources Law in the Faculty of Law at The University of Calgary. The Chair is a special senior position in the Faculty, established in 1978 with funding from the Canadian oil and gas industry. The Chair is designed to permit the Faculty to recruit outstanding scholars and teachers in natural resources law on either a short-term or long-term basis. Chair incumbents carry out research on natural resources issues and teach both in the Faculty and in programmes aimed at the legal profession and the public. The Chair receives significant funding from the Canadian Institute of Resources Law.

Professor Bradbrook is a leading authority in a number of fields of law, and has written texts and books in family, real property, residential tenancies, and solar energy law. During the course of his stay at The University of Calgary, his research was concentrated on the question of energy conservation. This volume represents part of the output of that research and addresses the legal problems that surround energy conservation, both with respect to new and existing buildings, and with respect to rental buildings. It includes not only a review of existing legislation, but also recommendations for possible legislative changes.

Professor Bradbrook is currently Professor of Law and Head of the University of Adelaide Law School.

J. Owen Saunders
Executive Director
Canadian Institute of Resources Law

AVANT-PROPOS

Ce livre présente la recherche effectuée par le professeur Bradbrook dans le cadre de ses fonctions en tant que président du droit des ressources naturelles de la Faculté de droit de l'Université de Calgary. Le poste de président a été créé en 1978 grâce à l'aide financière de l'industrie pétrolière et gazière du Canada. Il s'agit d'une fonction élevée dans la Faculté. Le président permet à la Faculté de recruter d'éminents étudiants et professeurs oeuvrant dans le domaine des ressources naturelles pour de courtes ou de longues durées. Les titulaires du titre de président font des recherches sur les questions relatives aux ressources naturelles et enseignent à la Faculté ainsi que dans le cadre de programmes destinés à la profession de droit et au public. Le président reçoit aussi une aide financière importante de la part de l'Institut canadien du droit des ressources.

Le professeur Bradbrook fait autorité dans un certain nombre de domaines en rapport avec le droit et a écrit des textes et des livres sur le droit de la famille, de la propriété, des loyers et de l'énergie solaire. À l'Université de Calgary, il a axé ses recherches sur la question de la conservation d'énergie. Ce livre présente une partie du fruit de cette recherche et fait état des problèmes juridiques concernant la conservation d'énergie des bâtiments nouveaux et existants ainsi que des bâtiments de location. Il comprend non seulement un rappel de la législation actuelle, mais aussi des recommandations pour des changements législatifs éventuels.

Le professeur Bradbrook est actuellement professeur de droit et directeur de l'école de droit de l'Université d'Adelaide.

J. Owen Saunders
Directeur exécutif
Institute canadien du droit
des ressources

ABSTRACT

This book provides a brief overview of the existing energy conservation legislation for new, existing, and rental buildings, and highlights the recommended legislative changes.

The existing legislation to encourage energy conservation in new and existing buildings includes financial incentives (stimulation) and building standards (regulation). The financial incentives legislation includes tax concessions, government-backed or government grants or loans to individuals and businesses, and statutory programs requiring energy utilities to improve energy efficiency in the community; in addition, specific funded programs apply to existing buildings.

The five types of existing energy conservation building standards for new buildings are mandatory building performance regulations, mandatory prescriptive regulations for building materials, mandatory or voluntary energy rating or labelling systems, a mandatory system of assigning energy and resource conservation option points to buildings, or a mandatory system requiring the vendor of the building to provide information in the contract of sale.

The application of energy conservation regulations to existing buildings is problematic because it is often difficult, if not impossible, to add such measures into buildings once they have been erected. A further difficulty is that it may be regarded as politically unacceptable to apply retrospectively new regulations to existing buildings.

The background policy issue that sets rental buildings apart from new and existing buildings as it relates to energy conservation measures is this: which party (landlord or tenant) has the duty to install and maintain energy conservation measures sufficient for the rental premises to meet a designated standard of energy efficiency? Further, although existing financial incentives (stimulation), building standards (regulation), and consumer protection legislation <u>appear</u> to encourage energy conservation in rental buildings, they do not do so.

The book concludes with a summary of the recommended legislative changes which would encourage home-owners, landlords, residential tenants, and commercial tenants to install energy conservation measures into new, existing, and rental buildings.

RÉSUMÉ

Ce livre présente un bref aperçu de la législation actuelle sur la conservation d'énergie des bâtiments nouveaux et existants ainsi que des bâtiments de location, et souligne les changements législatifs recommandés.

La législation actuelle mise en vigueur pour favoriser la conservation d'énergie des bâtiments nouveaux et existants comprend des encouragements financiers (mesures incitatives) ainsi que des normes de construction (règlements). La législation des encouragements financiers comporte des dégrèvements fiscaux, des garanties, des subventions ou des prêts des gouvernements pour les particuliers et les entreprises, et des programmes réglementaires selon lesquels les services publics du secteur de l'énergie doivent améliorer le rendement énergétique. De plus, des programmes financés sont mis en oeuvre pour les bâtiments existants.

Les cinq normes actuelles pour la conservation d'énergie des nouveaux bâtiments sont les règlements obligatoires pour le rendement des bâtiments, les règlements obligatoires normatifs pour les matériaux des bâtiments, l'estimation volontaire ou obligatoire de la consommation d'énergie ou le catalogage des systèmes, un système obligatoire pour attribuer des points d'option en matière de conservation d'énergie et de ressources pour les bâtiments, ou un système obligatoire selon lequel le vendeur doit fournir des renseignements sur le bâtiment dans le contrat de vente.

L'application des règlements sur la conservation d'énergie des bâtiments existants pose des problèmes parce qu'il est souvent difficile, sinon impossible, de mettre en place de telles mesures dans des bâtiments déjà construits. Et il pourrait être considéré inacceptable du point de vue politique d'appliquer ces mesures rétrospectivement aux bâtiments existants.

La question suivante distingue les nouveaux bâtiments de location des bâtiments nouveaux et existants en ce qui a trait aux mesures de conservation d'énergie : Qui, du propriétaire ou du locataire, a le devoir de mettre en place et de maintenir des mesures suffisantes de conservation d'énergie dans les bâtiments de location pour répondre aux normes définies de conservation d'énergie? De plus, bien que les encouragements financiers actuels (mesures incitatives), les normes de construction (règlements) et la législation sur la protection du consommateur *semblent* favoriser la conservation d'énergie dans les bâtiments de location, ce n'est pas le cas.

Ce livre se termine par un résumé des changements législatifs recommandés qui devraient encourager les propriétaires résidentiels et commerciaux de même que les locataires résidentiels et commerciaux à mettre en place des mesures de conservation d'énergie dans les nouveaux bâtiments, les bâtiments existants ainsi que dans les bâtiments de location.

1. INTRODUCTION

> Energy is not so much a single product as a mix of products and services, a mix upon which the welfare of individuals, the sustainable development of nations, and the life-supporting capabilities of the global ecosystem depend. In the past, this mix has been allowed to flow together haphazardly, the proportions dictated by the short-term pressures on and the short-term goals of governments, institutions, and companies. Energy is too important for its development to continue in such a random manner. A safe, environmentally sound, and economically viable energy pathway that will sustain human progress into the distant future is clearly imperative. It is also possible. But it will require new dimensions of political will and institutional co-operation to achieve it.[1]

This chapter notes the issue of energy conservation in buildings; examines whether governments should intervene to encourage energy conservation in buildings; defines "new", "existing", and "rental" buildings; and concludes with an overview of the law used by governments to intervene and to further the use of energy conservation measures in buildings.

As numerous studies have indicated,[2] a vast amount of energy is wasted to heat and cool unnecessary space due to the energy inefficient design and construction of buildings. A preliminary question to be addressed is whether government

---

1. World Commission on Environment and Development, *Our Common Future* (Melbourne, Aust.: Oxford University Press, 1990) [hereinafter *Brundtland Report*] at 246.
2. Energy conservation potentials of between 40 and 50% merely by modification of building regulations have been estimated. See, for example, G. Bergmann, R. Bruno & H. Horster, "Energy Conservation in Buildings" *in* J.F. Kreider & F. Kreith, eds., *Solar Energy Handbook* (New York: McGraw Hill, 1981) c.29; G.A. Lincoln, "Energy Conservation" *in* P.H. Abelson, ed., *Energy, Use, Conservation and Supply* (Washington: American Association for the Advancement of Science, 1974); *Energy, Global Prospects 1985-2000*, Report of the Workshop of Alternative Energy Strategies (New York: McGraw Hill, 1977).

intervention is necessary in building energy conservation.

Two opposing lines of economic thought regarding government intervention are considered here. One school of economic thought is that substantial energy savings may occur in due course by the operation of market forces, and that, therefore, government intervention is not necessary.[3] The argument has been made in the United States, in the context of a discussion of the need for domestic appliance efficiency legislation, that standards and financial incentives are unnecessary as the free market will produce equivalent savings. The argument runs that as energy prices increase, consumers will demand and manufacturers will produce more energy efficient appliances. The same argument would apply in respect of energy conservation devices in buildings.

However, the argument that government intervention is not necessary in building energy conservation was rejected by the California Energy Commission,[4] which conducted a survey which showed that the introduction of appliance efficiency standards was responsible for most of the improvements in appliance efficiency which have occurred in recent times in the United States. The Commission added that legislative efficiency standards *guarantee* that savings will occur and hasten the onset of the savings.[5]

Based on an analogy with domestic appliances, it is submitted that legislative action is required if significant energy savings in the building sector are to occur in the near future. In light of the seriousness of the environmental problems posed by global warming and the fact that energy use is the largest component of greenhouse gas emissions,[6] we cannot afford the

---

3. See, for example, A. Moran, "Sustainable Development or Reduced Economic Growth – Is There a Choice?", paper presented at Energy and the Environment Conference, Sydney, Australia, 9 April 1990.
4. California. California Energy Commission, *California's Appliance Standards: An Historical Review, Analysis and Recommendations* (Sacramento, Ca.: Report P400-83-020, 1983).
5. *Id.*, at 77-80. *See also* United States. General Accounting Office, *Appliance Efficiency Standards: Issues Needing Resolution by DOE* (Washington, D.C.: Report GAO/EMD-82-78, 1982) at 13.
6. Based on the U.S. Environmental Protection Agency estimates in 1989,

luxury of waiting to determine whether market forces will produce energy efficient buildings. Thus, the issue to be considered is not whether legislation is required, but rather, what forms of legislation should be enacted.

The book focuses on the law used by governments to intervene and to further the use of energy conservation measures in "new", "existing", and "rental" buildings. For the purposes of this book, "new" buildings means "never-sold, owner-used" buildings (either commercial or residential), as opposed to "existing" buildings (building sold at least once, but still owner-used) or "rental" buildings (commercial or residential).

In outline, for new and existing buildings, the existing legislation includes financial incentives (stimulation) and energy conservation building standards (regulation).[7] Financial incentives may be in the form of tax concessions, government-backed and government grants and loans, or statutory programs requiring energy utilities to improve energy efficiency in the community; the financial incentives would be included within the existing taxation legislation of each country. If building energy efficiency standards are adopted, they would most appropriately be included within the existing building standards legislation. Regulations ensure that new buildings reach a minimum prescribed threshold standard of energy performance, while stimulation can be devised so as to encourage builders and landowners to go beyond the minimum standard.

For rental buildings, the existing legislation includes consumer protection legislation, in addition to financial incentives and energy conservation building standards. Law reform of financial incentives and legislation governing

---

energy use and production is responsible for approximately 57% of greenhouse gas emissions in the United States. This figure may vary slightly between countries; the figures for third-world countries would be significantly less than for industrialized nations.

7. Financial incentives which take the form of government loans or grants may in many cases be implemented without the need for legislation, unless the receipt of the loans or grants have tax consequences. See, for example, *Income Tax Act*, S.C. 1970-71-72, c.63, s.12(1)(u), whereby recipients of now-discontinued Canadian federal government grants for insulation were required to declare such grants as income.

residential and commercial landlord-tenant relationships (landlord-tenant and general property legislation, respectively) is needed to clarify the respective duties of the parties in relation to making and keeping the premises energy efficient.

The relevant laws differs slightly between the common law jurisdictions, although the basic legal principles are similar. Examples of applicable legislation are chosen from the existing laws of a number of jurisdictions including Australia, Canada, Denmark, Japan, New Zealand, the United Kingdom, and the United States.

The book documents various types of "existing" building energy conservation legislation for both new and existing buildings (Chapter 2) and for rental buildings (Chapter 3). It surveys and analyzes all possible types of law reform, and proposes a "preferred" legal management model which could be adopted in all jurisdictions. The preferred legal management model may be a combination of a number of different means, or a modification of one or more means. The book concludes with a summary of recommendations.

2. ENERGY CONSERVATION LAWS FOR NEW AND EXISTING BUILDINGS

This chapter initially sets out, factually, the "existing" legislation used by various governments to promote energy efficiency in new buildings, noting any different requirements for existing buildings. A "preferred" legal management model to encourage energy efficiency in new and existing buildings is then proposed. The existing legislation is analyzed to determine its effectiveness, and some law reforms are suggested. The chapter concludes with a summary. The existing legislation includes financial incentives (stimulation) and energy conservation building standards (regulation).

Existing Financial Incentives Legislation

Significant financial incentives for the adoption of energy efficiency measures in building design and construction exist in Canada, the United States, and Japan. The financial incentives legislation examined includes tax concessions, government-backed or government grants or loans to individuals and businesses, and statutory programs requiring energy utilities to improve energy efficiency in the community. In addition to these, specific funded programs apply to existing buildings.

Tax Concessions

The means of granting tax concessions include tax write-offs, tax credits (both for residential and commercial buildings[1]), or the deduction of a percentage of the acquisition price of energy conservation equipment from income or corporate tax, or from capital cost allowance. Examples of each means will be given from specific jurisdictions.

The Canadian *Income Tax Regulations*[2] provide financial

---

1. For the purposes of this paper, "commercial" includes both "business" and "industry".
2. Enacted pursuant to the *Income Tax Act*, S.C. 1970-71-72, c.63.

incentives by means of tax write-offs for the installation of energy conservation measures in new commercial buildings. These incentives are contained in Class 34, Accelerated Capital Cost Allowance, of Schedule B of the Regulations,[3] and take the form of a 2 or 3-year tax write-off. Class 34, in conjunction with s.1100(1)(y) of the Regulations, allows eligible assets which conserve energy or use renewable sources of energy to be written off over two years, if acquired within the eligible period but before 12 November 1981 (up to 50 percent in the first year and the balance in any subsequent year), or over three years, if acquired within the eligible period but after 12 November 1981 (up to 25 percent in the first year, up to 50 percent in the second year, and the balance in any subsequent year). Without this accelerated write-off, the rates applied to these types of assets would normally be either 6 percent or 20 percent, calculated on a declining balance basis.

In the United States, Congress enacted income tax credits for energy conservation measures installed in new residential premises in Title I of the *Energy Tax Act* of 1978.[4] Section 44C of the Internal Revenue Code established, *inter alia*, an income tax credit for "qualified energy conservation expenditures".[5] The amount of the credit is specified as 15 percent of the energy conservation expenditures made by the taxpayer during the taxable year up to a maximum of $2,000.[6] The term "energy

---

3. Class 34 was originally added by P.C. 1976-2134, published in Canada Gaz., 22 September 1976, effective 25 May 1976.
4. Pub. L. No.95-618, 92 Stat. 3174-3205. For a discussion of this legislation *see* S.L. McDonald, "The Energy Tax Act of 1978" (1979) 19 Nat. Res. J. 859; S.E. Ferrey, "Solar Banking: Constructing New Solutions to the Urban Energy Crisis" (1981) 18 Harv. J. Legis. 483 at 485-488; S.F. Knopf, "Tax Benefits Through the Use of Solar Energy" (1980) 2 Northrop U.L.J. of Aerospace, Energy and the Environment 85; J.H. Minan & W.H. Lawrence, "State and Federal Tax Incentives to Promote Solar Use" *in* J.H. Minan & W.H. Lawrence, eds., *Legal Aspects of Solar Energy* (Massachusetts: Lexington Books, 1981) c.4; S.F. Kraemer, *Solar Law* (Colorado Springs: Shephard's, 1978), App.4.
5. This section also provided an income tax credit for "renewable energy source expenditures".
6. 26 U.S.C. § 44C(b)(1). Between 1978 and 1983, 24 million households

conservation expenditures" is defined as an expenditure made on or after 20 April 1977 by the taxpayer for insulation[7] or any other energy-conserving component installed in or on a dwelling unit used by the taxpayer as their principal residence.[8] The component must reasonably be expected to remain in operation for at least five years and must meet the performance and quality standards (if any) prescribed by the regulations. Section 44C was stated not to apply to expenditures made after 31 December 1985.[9] A concerted campaign by lobbyists during 1984 and 1985 to remove this sunset clause was vigorously opposed by the Reagan administration and was ultimately unsuccessful.[10] Although a similar business investment tax credit was established in respect of renewable energy expenditures,[11] it did not extend to energy conservation expenditures.

In addition to the income tax credits offered by the U.S. Congress, a number of states of the United States have enacted various income tax incentives for the installation of renewable energy devices and energy conservation measures in new residential premises, and the majority give some form of business tax credit.[12] In most instances, the state legislation

---

claimed a residential energy conservation tax credit at a total cost in terms of lost revenue to the U.S. federal treasury of $2.3 billion: American Council for an Energy Efficient Economy, *Energy Efficiency in Buildings — Progress and Promise* (Washington, D.C.: 1986) at 157.

7. "Insulation" is defined in 26 U.S.C. § 44C(c)(3) to mean:

   any item —
   (A) which is specifically and primarily designed to reduce when installed in or on a dwelling (or water heater) the heat loss or gain of such dwelling (or water heater);
   (B) the original use of which begins with the taxpayer;
   (C) which can reasonably be expected to remain in operation for at least 3 years; and
   (D) which meets the performance and quality standards (if any) which —
   i) have been prescribed by the Secretary by regulations; and
   ii) are in effect at the time of the acquisition of the item.

8. 26 U.S.C. § 44C(c)(1).
9. 26 U.S.C. § 44C(f).
10. See, for example, Note, "Federal Tax Credits Expire" (1986) 7(1) Wisconsin Energy News 5.
11. Internal Revenue Code, 26 U.S.C., §§ 46, 48.
12. California. California Energy Commission, *Solar and Wind Technology: Tax Incentive Impact Analysis* (Sacramento, Ca.: Report P500-86-010,

permits the taxpayer to claim both the federal and the state credits simultaneously. Many state tax credits, like the U.S. federal tax credits, have expired pursuant to a sunset clause, although in a significant number of states the credits still remain in effect. A noteworthy feature of the state legislation is that in most instances there is a maximum dollar limit specified in respect of the tax credit available, with the result that a greater incentive is provided for smaller, less expensive energy conservation measures than for larger, more expensive measures.[13]

Under Japanese legislation, if energy conservation equipment installed in commercial buildings is put into use within one year after its acquisition, then in the following year the investor can either deduct 7 percent of the standard acquisition price of the equipment from business income or corporation tax (up to a maximum of 20 percent of the tax assessed), or claim under the capital cost allowance at a special rate of up to 30 percent of the standard acquisition price in addition to the ordinary rate. If the equipment is imported, investors are also entitled to a 20 percent increase in the tax reduction rate (or special depreciation rate) under the system. In addition, the standard value of municipal property tax on energy conservation equipment installed in buildings can be reduced to 5/6 for 3 years.[14]

Financial incentives for energy conservation certainly need not be restricted to tax concessions. Another method of encouraging the use of energy conservation measures is to ensure that government-backed or government grants or loans for the purchase of these types of devices and measures are readily available.

1986) at 2-7.

13. In Arizona, California, Hawaii, North Dakota and Oklahoma, there is no maximum limit specified.

14. Japan. Energy Conservation Center, *Energy Conservation in Japan* (Tokyo: Energy Conservation Center, 1988) at 27.

Government-backed or Government
Grants or Loans

Studies in the United States have shown that banks have been reluctant to make substantial housing loans for the purchase and installation of energy conservation measures, and that they frequently exclude the value of such measures when determining the appraised value of a house to determine the maximum amount of money they are prepared to loan.[15] Government-backed or government grants and loans can overcome this problem.

Government-backed grants or loans can be distributed by authorized utilities or other agencies, or by banks designated to back energy conservation measures; government grants can be given directly to industries which manufacture energy conservation materials and equipment, or to individuals for renewable energy and conservation purposes. Examples will be given from specific jurisdictions.

Various electric utilities in the United States have adopted loan-making schemes. These schemes vary greatly.[16] In many cases, the schemes were adopted reluctantly by the electric utilities under pressure from the State Public Utilities Commission.[17] In Australia, the Victorian government has issued an Energy Policy Statement on its Renewable Energy Strategy,[18] which announced a financial incentives program for loans and grants by the gas and electric utilities for the purchase of certain types of renewable energy devices and energy conservation measures. Suitably expanded in scope, this program

---

15. S.E. Ferrey, "Solar Banking: Constructing New Solutions to the Urban Energy Crisis" (1981) 18 Harv. J. Legis. 483 at 512.
16. *See* W.H. Lawrence & J.H. Minan, "Financing Solar Energy Development Through Public Utilities" (1982) 50 Geo. Wash. L. Rev. 371 at 390-394; E. Kahn, "Using Utilities to Finance the Solar Transition" (1980) 2 Solar Law Reporter 543; M.C. Gardels, "Utility Financing: Loans or Rebates?" (1981) 3 Solar Law Reporter 477.
17. See generally B.R. Barkovich, *Regulatory Interventionism in the Utility Industry* (New York: Quorum Books, 1989).
18. Victoria. Dept. of Industry, Technology and Resources, *Victorian Energy Planning Program: Government Energy Policy Statement — Renewable Energy Strategy* (Melbourne, Aust.: 1985) at 8.

could be the forerunner of an effective widespread incentives scheme to promote all types of energy conservation measures.

Other agencies operate government-backed loan schemes in the United States at the state level. In Texas, money raised in oil overcharge litigation[19] is administered by the State Energy Management Center for various purposes, including loans and grants for the purchase of energy conservation equipment.[20] Individuals who purchase a house with low interest mortgage money from the South Dakota Housing Development Authority are eligible to receive assistance in making efficiency improvements to their home.[21] In California, loans are provided through numerous state agencies and utilities.[22]

Perhaps the most radical and far-reaching initiative taken to encourage energy conservation was the creation by President Carter in the United States of the Solar Energy and Energy Conservation Bank.[23] This Bank was established by Title V of the *Energy Security Act* of 1980,[24] which provided for a range of flexible subsidies to lending institutions designed to overcome barriers to private individuals and corporations obtaining access to credit on reasonable terms for investment in renewable energy devices and energy conservation measures. The legislation allowed the Bank to transact business with lending institutions in

---

19. *See Oil Overcharge Restitutionary Act* (Art.4413(56), Vernon's Texas Civil Statutes). This source of money results from the settlement made between the U.S. federal government and various oil companies for violations of price controls on crude oil and petroleum products that were in effect during the 1970s.
20. Texas Government Code § 447.001*ff.*
21. South Dakota. Energy Office, *Energy Savings and Cost-Effectiveness in the South Dakota Housing Development Authority Energy-Efficiency Program* (1986) v.1, at 3.
22. *See* California. California Energy Commission, *A Compendium of Major California Conservation Programs* (Sacramento, Ca.: Report P300-84-017, 1984).
23. The Solar Energy and Energy Conservation Bank is discussed in Ferrey, *supra*, note 15; C.L. Deering, "Solar Energy and Energy Conservation Bank" *in* L.E. Buck & L.M. Goodwin, eds., *Alternative Energy — The Federal Role* (New York: McGraw-Hill, 1982) c.3.
24. *Energy Security Act 1980*, Pub. L. No.96-294, 94 Stat. 611, codified in 12 U.S.C. §§ 3601-3620.

two ways. First, the Bank was permitted to make lump-sum payments to the lending institutions. These payments were used either to make conservation grants to eligible recipients or to reduce the principal or interest on solar energy or energy conservation loans. Secondly, the Bank was permitted to engage in secondary-market loan financing by purchasing any loan made wholly or partly for renewable energy or energy conservation purposes.[25] The Bank was not, however, permitted to make direct loans to consumers, so its title was misleading. When initially established, the Bank was appropriated $2.5 billion for energy conservation and $525 million for renewable energy purposes for the period 1980-1984.[26] Pursuant to a sunset clause, the Bank ceased to exist on 15 March 1988,[27] but throughout its lifetime, it was harassed and made less effective than its proponents envisaged by the actions of a hostile Reagan administration.[28]

Japanese banking provisions are similar to, but not identical to, those of the United States. In Japan, loan programs for energy efficiency are administered by the Japan Development Bank, the Small Business Finance Corporation, the National Finance Corporation, the Hokkaido/Tohoku Development Finance Corporation (specifically for cogeneration), and the Housing Finance Corporation.

The use of direct government grants or loans for industries not only encourages the establishment of relevant industries, but has the effect of reducing the sale price of merchandise to consumers. In Australia, the Victorian government has given grants to particular industries for many years to assist them to commercialize renewable energy technologies.[29] Such measures,

---

25. Ferrey, *supra*, note 15, at 524.
26. 12 U.S.C. § 3620.
27. 12 U.S.C. § 3603.
28. *See* B. Stambler, "Solar Bank Victory", *Sun Times*, July/August 1982, 12; Note, "President Recommends Abolishing Solar Bank, Cutting Solar Budget" (1981) 2 Solar Law Reporter 1032; Note, "Two Groups Lodge Impoundment Suit Over Solar Bank Funds" (1982) 3 Solar Law Reporter 741; Note, "Solar Bank Victory" (1982) 8(9) Northern California Sun 5.
29. Victoria. Dept. of Industry, Technology and Resources, *Victorian*

suitably expanded in scope to include energy conservation technologies, would be of great assistance to fledgling industries if adopted in any country.

The use of direct government grants to individuals for energy conservation purposes has been adopted in Canada, the United Kingdom, and Japan. In Canada, direct grants were made to private individuals by the federal government under the terms of the *Canadian Home Insulation Program Act* and accompanying regulations.[30] Pursuant to s.3(1), the legislation applied only to persons owning or leasing a family housing unit in a residential building of not more than three storeys in height, and residential buildings of a similar height owned and operated by a non-profit corporation, a department or an agency of the Government of Canada or of a province, a municipality or an agency thereof, or an institution providing post-secondary education. By the terms of s.3(2), the legislation expired on 31 December 1987.

In the United Kingdom, under the now-discontinued *Homes Insulation Act 1978*, s.1, local authorities were empowered to make grants to private individuals towards the cost of works undertaken to improve the thermal insulation of dwellings in their areas. The legislation permitted the Secretary of State to specify from time to time the insulation works for which grants were available, the description of dwellings which qualified, and the persons from whom applications might be entertained (s.1(2)). Under the legislation, the availability of grants could be limited to those applying on grounds of special needs, such as age, disability, bad health and inability to finance the cost of the works (s.l(5)(6)). Finance for the making of grants was provided to local authorities from time to time by the Secretary of State (s.2(l)).

The British scheme has been replaced by the *Home Energy Efficiency Grants Regulations 1990*,[31] made pursuant to s.15 of

---

Energy Planning Program: Government Energy Policy Statement – Renewable Energy Strategy (Melbourne, Aust.: 1985) at 8.

30. *Canadian Home Insulation Program Act*, S.C. 1980-81-82-83, c.57; SOR/81-437, 29 May 1981.

31. *Home Energy Efficiency Grants Regulations 1990*, S.I. 1990/1791

the *Social Security Act 1990* (U.K.). The scheme is much narrower than that under the 1978 legislation as, pursuant to Regulation 3 of the 1990 Regulations, applications for a grant may only be entertained by resident householders where either the householder or their spouse is, at the time of making the application, in receipt of any income support, housing benefit, family credit, or community charge benefit (within the meaning of s.20 of the *Social Security Act 1986*). Grants are limited to work which provides roof space insulation, draughtproofing, insulation of hot water tanks or cylinders, and energy advice (Regulation 6(1)). The work must comply with such standards, including standards of materials and workmanship, as are laid down from time to time by the administering agency with the approval of the Secretary of State (Regulation 6(2)).

The Japanese government directly subsidizes the installation of solar energy systems in public buildings, and provides low interest loans for the installation of such systems in houses and offices.[32]

Statutory Utility Programs to Improve Energy Efficiency

As an alternative to tax concessions, or to government-backed or government loans or grants programs, the energy utilities may be required by legislation to undertake programs to improve energy efficiency in the community. The financial incentive to the utility for such improvements is that it will not need to incur debt to construct other power stations. Under the *National Energy Conservation Policy Act 1978*, utilities in the United States are required to offer energy audits to residential customers and must offer to arrange for the installation and financing of measures identified.[33] The states are encouraged by the federal government to include the utilities in their Energy Plans. Perhaps the best example is provided by the Pacific Northwest Power and Conservation Planning Council, which requires the local utilities to educate people in energy

---

(U.K.).

32. Japan Energy Conservation Center, *supra*, note 14, at 27-28.

33. *National Energy Conservation Policy Act 1978*, 26 U.S.C. § 8213.

management, to conduct conservation programs, and to provide specific minimum levels of financial assistance to people who satisfy the Model Conservation Standards set down by the Council. If a utility fails to achieve the requisite level of energy saving, the Bonneville Power Administration is empowered to impose a surcharge on it.[34]

Specific Funded Programs for Existing Buildings

The alternative systems of government-backed or government grants or loans, tax incentives, or statutory utility programs to improve energy efficiency, discussed above in the context of new buildings,[35] also apply to existing buildings. In addition, the majority of common law jurisdictions which have legislated to improve energy conservation in existing buildings have adopted specific funded programs. This has occurred in the United Kingdom, where the *Home Energy Efficiency Grants Regulations 1990* provide for a system of local authority grants towards the thermal insulation of dwellings.[36] The former has occurred in the United States, where under s.451(a)(1) of the *Energy Conservation in Existing Buildings Act 1976,*[37] the Administrator is enabled to guarantee loans entered into by any person for the purpose of financing any energy conservation measure to be installed in any building or industrial plant. The program operates in conjunction with weatherization programs in effect in a number of states (for example, California and South Dakota) whereby specified conservation measures are installed in residential dwellings at a reduced cost or free of charge after an official energy audit of the premises conducted by a government body.

---

34. 51 Fed. Reg. 7364 (1986). *See also Pacific Northwest Electric Power Planning and Conservation Act 1980*, 16 U.S.C. §§ 839-939h.
35. See *supra*, notes 1-31 and accompanying text.
36. See *supra*, note 31 and accompanying text.
37. *Energy Conservation in Existing Buildings Act 1976*, Pub. L. 94-385, 90 Stat. 1150. *See also* §§ 413, 414, and 432 of this Act for other energy conservation measures adopted in the U.S. in respect of existing buildings: for example, weatherization program, financial assistance, and demonstration programs.

While financial incentives legislation encourages the design and construction of energy efficient measures in new buildings, energy conservation building standards are needed to control the design and construction of energy efficient measures in new buildings. As the installation of energy conservation measures into new and existing buildings differs considerably, the author will first discuss the application of energy conservation building standards for new buildings, and then for existing buildings.

Existing Energy Conservation Building Standards Legislation for New Buildings

The majority of jurisdictions possess comprehensive building standards regulations or by-laws, which codify the existing law relating to the design and construction of buildings. While some countries' building standards regulations contain a variety of energy conservation measures, other countries have no such measures, including many western industrialized countries. For example, in Australia, there are no energy conservation requirements specified in the Building Code of Australia or any of the states' building regulations,[38] the only exception being Victoria, which, in 1991, introduced mandatory insulation requirements for private-dwellings into the Victorian Building Regulations.

Among those countries which have addressed the issue of energy conservation in their building standards laws, five different types of energy conservation building standards for new buildings can be identified: mandatory building performance regulations, mandatory prescriptive regulations for building materials, mandatory or voluntary energy rating or labelling

---

38. The states' building regulations are contained in the *Victoria Building Regulations 1983*, made pursuant to the *Building Control Act 1981* (Vic.); Ordinance No.70, "Building", made pursuant to the *Local Government Act 1919* (N.S.W.); Standard Building By-laws 1975, made pursuant to the *Building Act 1975* (Qld.); *Building Regulations 1973*, made pursuant to the *Building Act 1971* (S.A.); *Building Regulations 1989*, made pursuant to the *Local Government Act 1960* (S.A.); *Building Regulations 1978*, made pursuant to the *Local Government Act 1962* (Tas.).

system for buildings, a mandatory system of assigning energy and resource conservation option points to buildings, or a mandatory system requiring the vendor of the building to provide information on the insulation installed in the building in the contract of sale. Examples of the five different types of energy conservation building standards will be given from specific jurisdictions.

Building Performance Regulations and Prescriptive Regulations for Building Materials

The first type of energy conservation building standard for new buildings, building performance regulations, sets energy conservation criteria for the building as a whole.[39] Under building performance regulations, considerable flexibility is given to builders and architects in that they are free to use any types of designs and materials, provided that the building as a whole conforms to certain specified minimum of energy performance. Thus, for example, in relation to thermal insulation, a building performance code will not prescribe minimum levels of insulation, but will specify the maximum level of heat loss permissible from the building. In contrast, the second type of energy conservation building standard for new buildings, prescriptive regulations for building materials, specifies the minimum acceptable levels of designated materials which are required to satisfy the regulations, and give builders and architects no scope for flexibility in the use of building materials. It is possible to incorporate elements of both systems into any building regulations, and to allow builders to elect between the application of either system in respect of each building.

---

39. *See* W.D. Kopper, "Energy Building Regulations: The Effect of the Federal Performance Standards on Building Code Administration and the Conservation of Energy in New Buildings" (1980) 13 U. Calif. at Davis L. Rev. 330 at 335*ff*; C.W. Bullard, "Energy Conservation in New Buildings: Opportunities, Experience, and Options" (1981) 6 Ann. Rev. Energy 199; G.P. Thompson, *Building to Save Energy: Legal and Regulatory Approaches* (Cambridge, Mass.: Ballinger Publishing Co., 1980) at 33*ff*.

Canada

In Canada, for constitutional reasons, energy conservation building standards are primarily within the jurisdiction of the provinces. The federal government is only able to apply building regulations to buildings owned or financed by the Government of Canada.[40]

Mandatory building performance standards were promulgated in 1978 by the Canadian federal government in the document entitled "Measures for the Conservation of Energy in New Buildings".[41] These standards were originally developed in the United States by the American Society of Heating Refrigerating and Air-conditioning Engineers (ASHRAE),[42] and were modified by the Canadian federal government to suit Canadian climatic conditions.[43] In place of traditional insulation standards, the 1978 standards specify minimum thermal resistance values for each part of a building, including wall assemblies above ground level, foundation wall assemblies, floor assemblies, perimeters of slab-on-ground floors, glazing, and specifies maximum standards of airtightness. The minimum thermal resistance standards vary according to the maximum number of degree days recorded by Environment Canada for each municipality, and vary according to the use of the building. Thus, for example, more stringent minimum thermal resistance standards apply to private houses than to commercial buildings. The Canadian federal government building standards were adopted shortly afterwards by the Quebec legislature, which legislated them into effect in 1980 in the *Conservation of Energy in Buildings Act*.[44] This Act applies not only to private houses, but also to a "public building", which is defined in part by s.1 as

---

40. *See* Canada. Dept. of Energy, Mines and Resources, *Energy Use and Atmospheric Change: A Discussion Paper* (Ottawa: 1990) at 39-40.
41. Canada. National Research Council, "Measures for the Conservation of Energy in New Buildings", 2d ed. (Ottawa: N.R.C. Document No.22432, 1983).
42. See *infra*, notes 52*ff* and accompanying text.
43. Heavy reliance is placed on ASHRAE Standard 90A-80 (Energy Conservation in New Building Design).
44. *Conservation of Energy in Buildings Act*, S.Q. 1980, c.32.

"a building used for the production or distribution of goods or services". This Act gives the Quebec government very broad powers to prescribe by regulation different types of energy efficiency and conservation measures for new buildings, including minimum levels of energy efficiency for products and construction methods.[45]

In the remainder of Canada, energy conservation building standards are based solely on prescriptive standards for building materials. Each province except Quebec has adopted, subject to certain amendments, the National Building Code of Canada, 1985[46] as its provincial Building Code.[47] In respect of energy conservation building standards, the National Building Code contains only prescriptive standards for insulation, which are confined in their operation to houses and small buildings up to 600 square metres per floor and three storeys in height (assembly, institutional, and high hazard buildings are exempted). In respect of other buildings, the only requirements specified are those necessary to avoid moisture condensation. With regard to houses and small buildings, the National Building Code of Canada adopts various insulation standards prepared by the Canadian Standards Association and the Canadian General Standards Board.[48] The prescribed insulation standards vary according to a fixed scale based on the number of degree heating

---

45. The relevant regulation is the Quebec *Regulation Respecting Energy Conservation in New Buildings*, O.C. 89-83, 19 January 1983, G.O.Q. 1983.II.775.
46. Canada. National Research Council, *National Building Code of Canada, 1985* (Ottawa: N.R.C. Document No.23194, 1985).
47. See, for example, *Alberta Building Regulation*, Alta. Reg. 186/85, made pursuant to the *Uniform Building Standards Act*, R.S.A. 1980, c.U-4.
48. The following Canadian Standards Association (CSA) and Canadian General Standards Board (CGSB) standards are adopted by para.9.26.3.3 of the *National Building Code of Canada, 1985*: CSA A101 (Thermal Insulation, Mineral Fibre, for Buildings), CSA A247 (Insulating Fibreboard), CGSB 51-GP-20M (Thermal Insulation, Expanded Polystyrene), CGSB 51-GP-21M (Thermal Insulation, Urethane and Isocyanurate, Unfaced), CGSB 51-GP-25M (Thermal Insulation, Phenolic, Faced), CGSB 51-GP-27M (Thermal Insulation, Polystyrene, Loose Fill), and CSGB 51-GP-60M (Thermal Insulation, Cellulose Fibre, Loose Fill).

days[49] which apply in the relevant locality. Thus, houses located in severe climatic areas are required to have a higher level insulation than those in milder areas. Unfortunately, the maximum level of insulation prescribed in the adopted standards is unrealistically low, being R-20 for walls and R-32 for ceilings of new houses. This standard is adequate in the mildest parts of Canada, but is inadequate for most of the country. In Alberta, for example, the maximum R-20 and R-32 values apply to both Calgary and Edmonton,[50] even though the building industry recommends R-25 and R-40 values for walls and ceilings in Edmonton, and even higher values for more northerly localities.[51]

United States

In the United States, energy conservation building standards for new buildings have been devised by both the American Society of Heating Refrigerating and Air-conditioning Engineers (ASHRAE)[52] and the Building Officials and Code Administrations International Inc. (BOCA). The ASHRAE and BOCA standards represent a combination of both prescriptive and building performance regulations.

BOCA has published its standards in its Basic National Model Building Code.[53] The intent of the ASHRAE and BOCA documents is to attempt to specify in detail minimum requirements for energy efficient building design, to regulate the

---

49. "Degree heating days" means the total number of degrees between 18 C and the mean daily temperature for each day in the year when the mean daily temperature is below 18 C.
50. The "R-value" is a measure of the thermal resistance and thus, of the effectiveness of insulation. The higher the R-value, the greater the power of insulation. In each case, the R-value will depend on the density, thickness, and type of insulating power.
51. Information provided by Dr. T. Lee, Department of Environmental Design, University of Calgary, February 1991.
52. ASHRAE Standard 90A-80 (Energy Conservation in New Building Design); ASHRAE Standard 90.1P (Energy Efficiency Design of New Buildings Except Low-Rise Residential Buildings).
53. Building Officials and Code Administrations International, Inc., *The BOCA Basic National Building Code*, 9th ed. (Washington, D.C.: 1984).

design of building envelopes and construction for adequate thermal resistance and low air leakage, and to control the design and selection of mechanical, electrical, and illumination systems and equipment so as to promote and provide the efficient use of energy in new buildings. The content of the standards includes:

1. the establishment of categories of buildings and the creation of different standards for different categories;
2. the exclusion of certain types of buildings from the standard;
3. variations in the energy conservation building standards within the same category of buildings to take account of climatic variations;
4. thermal resistance standards;
5. minimum heat loss standards for openings in the outer shell of a building;
6. quality standards and installation methods for insulation materials;
7. measures for preventing condensation in buildings;
8. standards for air or water heating or cooling systems, for lighting and ventilation systems, and for the related control devices; and
9. fenestration standards.

Mandatory energy conservation building standards have been enacted in the United States by all levels of government. At the federal level, the *Energy Conservation and Production Act* of 1976, Title III, required the Department of Housing and Urban Development to develop, promulgate, implement, and enforce compliance with building performance standards to improve the energy efficiency of all new buildings in the nation. By a 1977 amendment, responsibility to develop and promulgate the standards passed to the Department of Energy. The law provides that the standards will be voluntary for new non-federal buildings, but mandatory for new federal buildings. In January 1989, the U.S. Department of Energy promulgated interim energy conservation performance standards for new commercial and multi-family high rise residential dwellings. The interim standards do not regulate the design of non-federal buildings, although the Department of Energy recommends the use of the standards as guidelines for the design of energy conserving buildings. The format of the interim standards is similar, but not identical, to that proposed in ASHRAE Standard 90.1P, recommended for the design of new commercial buildings.

The majority of the U.S. states have now enacted building energy conservation codes, based on either the ASHRAE or the BOCA models.[54] In some cases, the legislation simply enacts the ASHRAE or BOCA standards into law, while in other cases minor variations have been made to make the standards relevant for local climatic conditions or for other reasons.

A few states have adopted different approaches. For example, the California Building Efficiency Standards are similar to the ASHRAE codes, but buildings are divided into six categories and specific sets of standards are set out for each, with all buildings required to have some form of insulation installed.[55] The Pacific Northwest Electric Power and Conservation Planning Council has developed Model Conservation Standards which impose standards of envelope performance on new residential buildings and adopt the BOCA Code only in relation to new commercial buildings.[56] In Maine, the *Energy Efficiency Building Performance Standards Act* sets out simplified standards for envelope thermal performance for one- or two-family residential structures, and requires multi-family residential, commercial, and residential buildings to conform with the ASHRAE standard for new building design. Commercial and institutional buildings need not comply with these standards if they are constructed in anticipation of sale or rental, and are so constructed that a buyer or lessee will be able to readily install any energy conservation measures required for any potential uses of the building.[57] Vermont legislation states

---

54. See, for example: Ct. State Building Code, Art.24 (Energy Conservation); Ind. Administrative Code, Title 675 (Energy Conservation); Ky. Rev. Stat., s.198B.050; Ky. Building Code, Art.31 (Energy Conservation); Mich. State Construction Code, Public Act 230 of 1972; NH Code for Energy Conservation in New Building Construction (3d ed., 1986); Nev. Administrative Code, c.523 (Conservation of Energy in Buildings).
55. Calif. Code of Regulations, Title 24, Part 2, c.2-53.
56. *See* 51 Fed. Reg. 7365 and 7369 (1986).
57. Me. Stats., Title 10, c.214.

that a residential building permit will be granted where the plans reflect the principles of energy conservation and incorporate the best available technology for efficient use of energy.[58] The Vermont Department of Public Service has interpreted this to mean either the least energy use, or the lowest life cycle cost, and has set out guidelines which are a simplified version of the state standards.[59]

Most of the state legislation in the United States applies state-wide, although it is usually enforced at the local government level. In some cases, the state legislation permits local municipalities with more stringent local laws to opt out of the state law. Davis, California, has enacted energy conservation standards in its residential building code,[60] which are more rigorous and wide-ranging than those generally applicable in California. The California Energy Commission has certified Davis' residential building code in lieu of the state's Energy Conservation Building Standards. For new residential buildings, the Davis regulations cover the state standards, and go on to include standards relating to the siting and orientation of buildings on blocks of land, shading, landscaping and subdivision.

Elsewhere in the United States, the U.S. Congress and some state legislatures (for example, Utah) have adopted mandatory energy conservation building standards, based on either the BOCA or ASHRAE codes, which are restricted in their operation to federal or state government buildings.[61] These jurisdictions

---

58. Vt. Act 250 of 1985.
59. *See* Vermont. Public Services Dept., Conservation and Renewable Energy Unit and Planning Division, *Guide to Evaluating Energy Conservation Opportunities* (1986); Vermont. Dept. of Public Services, *Energy Efficiency Recommendations for Residential Development* (1989).
60. City of Davis, Calif., Ordinance No.1049 (7 November 1979), which adds Division 4 (Energy Conservation Retrofit Regulations) to the City of Davis Code 1971. Further energy conservation requirements were added by Ordinance No.1060 (1981).
61. Energy Conservation Voluntary Performance Standards for New Commercial and Multi-Family High Rise Residential Buildings; Mandatory for New Federal Buildings, 54 Fed. Reg. 4538 (1989); 42

also allow, encourage, or require local municipalities to develop their own local standards.

New Zealand

Minimum prescriptive insulation requirements for dwellings exist in New Zealand, and are implemented at the local government level. Section 644(2) of the *Local Government Act 1974* (N.Z.), as amended in 1979, states:

> (2) The [municipal] council shall, for the purpose of preventing or reducing heat losses in residential buildings as an energy conservation measure, make by-laws, which shall apply to new buildings only, for the following purposes:
> (a) Specifying the minimum thermal requirements for residential buildings.
> (b) Specifying the manner in which those requirements are to be met.
> (c) Requiring owners of residential buildings to comply with those requirements.

The empowering bylaw is contained in the local bylaw, New Zealand Standards 1900, s.4.26:

> 4.26.1 Every residential building shall have thermal insulation properties approved as adequate to prevent an excessive requirement for space heating that utilizes energy from a public electricity supply, from any fuel, or from any other depletable energy source.
>
> 4.26.2 A residential building that complies with the requirements of NZS 4218P[62] in accordance with NZS 4214 shall be approved as complying with the requirements of 4.26.1.[63]

The New Zealand government proposes that this legislation be replaced by new legislation which will specify energy conservation performance standards for both residential and

---

U.S.C.A., § 8253; Utah Code Ann, s.63-9-66.

62. NZS 4218P (adopted in 1977) lays down minimum insulation requirements for residential buildings and is approved as a means of compliance with the relevant requirements of NZS 1900, c.4. The New Zealand government proposes to revise this statute.

63. NZS 4214 (adopted in 1977) sets out methods for determining the total thermal resistance of parts of buildings in steady-state environmental conditions. It is approved as a means of compliance with the relevant requirements of NZS 1900. The New Zealand government proposes to revise this statute.

commercial buildings.[64]

United Kingdom

Energy conservation building standards have had a chequered history in the United Kingdom. The relevant legislation is the *Building Act 1984*. Standards for the conservation of fuel and power were initially made under this Act by the *Building Regulations 1985*, Schedule 1, Part L.[65] This Schedule provided two alternative means of compliance: prescriptive insulation standards, or new building performance calculation procedures. As under the previous law, standards were prescribed relating to the insulation of walls, ceilings and floors, glazing, the insulation of pipes and ducts, and control of space and water heating. In addition, new building performance calculation procedures allowed variations in the insulation levels for individual elements within a permitted rate of heat loss for the whole building and a more sophisticated system of energy targets to allow greater flexibility.

Schedule 1, Part L, was repealed by the *Building Regulations (Amendment) Regulations 1989*,[66] and replaced with the following requirement: "Reasonable provision shall be made for the conservation of fuel and power in buildings". The requirement is stated to apply to dwellings and other buildings whose floor area exceeds 30 square metres. Section 6 of the *Building Act 1984* empowers the Secretary of State to approve and issue documents which provide practical guidance on meeting the requirements in the regulation.

An Approved Document has been promulgated on "Conservation of Fuel and Power", pursuant to the new Part L of Schedule 1. It contains detailed specifications relating to limitation of heat loss through the building fabric, controls for space heating and hot water supply systems, and insulation of

---

64. Information supplied by the Hon. David Butcher, New Zealand Minister of Energy, 24 July 1990. The performance-oriented energy efficiency standards will be incorporated into the proposed Buildings Bill 1992.
65. *Building Regulations 1985*, Schedule 1, Part L, S.I. 1985/1065 (U.K.).
66. *Building Regulations (Amendment) Regulations 1989*, S.I. 1989/1119 (U.K.).

hot water storage vessels, pipes, and ducts. Under the new British system, compliance with an Approved Document is not mandatory, as such documents are designed merely to provide standard ways of meeting the legal requirements. If the Approved Document is followed, the enforcing authority will accept that the building work complies with the legal requirements. If it is not followed, the onus is on the builder to satisfy the enforcing authority that their alternative solution is satisfactory. The system has the advantage of avoiding lengthy, complex building requirements being included with the Regulations, although it still requires the details to be included in other legal documents. Unfortunately, however, the wording of the new Schedule 1, Part L, is vague and lacks specificity. Indeed, where the Approved Document is not adhered to, Schedule 1, Part L, is so vague that it appears to be incapable of enforcement by the appropriate building authorities.

Other Countries

Within Europe, most member states of the European Economic Community have developed building performance standards for the allowed heat transmission losses of a building; these vary depending on the building or the climate.[67] France has examined the standards for the efficiency of heating plant and the energy required to heat a building.[68]

A performance-based system also operates in Japan. The Minister of Construction has been empowered to prepare and announce standards for the prevention of heat loss through the walls of a building and the effective use of energy in air-conditioning equipment in buildings to be constructed, enlarged, or reconstructed. The Minister may give guidance and advice to owners of non-residential buildings, and may announce guidelines, based on the standards, for residential buildings. The Minister may also give guidance and advice to manufacturers of

---

67. Commission of the European Communities, Directorate-General for Energy, *Thermal Insulation and Ventilation in Buildings — A Comparison of Requirements in EEC-Member States* (Brussels: Commission of the European Communities, 1987).
68. *Id.*, at 59*ff.*

insulating materials.[69]

Energy Rating or Labelling System

The third type of energy conservation building standard for new buildings is a mandatory or voluntary energy rating or labelling system for new buildings.[70] Whereas mandatory energy labelling is widely found in respect of domestic appliances under legislation and regulations enacted by common law jurisdictions,[71] mandatory energy rating or labelling systems for new buildings only exist in a few jurisdictions. Such a system has recently been adopted by the California Energy Commission.[72] Where they exist, energy rating or labelling systems for buildings perform several functions: to encourage builders to select energy-efficient properties, to induce home buyers to select energy-efficient homes, to encourage home owners to improve the energy efficiency of their property, and to facilitate the inclusion of energy efficiency in appraisals and lending decisions by mortgage lenders.[73]

---

69. Japan Energy Conservation Center, *supra*, note 14, at 25.
70. *See* A.H. Rosenfeld & L. Schuck, "Home Energy Ratings: A Promising Approach to Energy Conservation", (September 1984) Critical Mass Bulletin 3.
71. Canada: *Consumer Packaging and Labelling Act*, R.S.C. 1985, c.C-38, s.18. United States: *Energy Policy and Conservation Act 1975*, Pub. L. No.94-163, 89 Stat. 871. Australia: *Energy Labelling and Standards (Refrigerators, Freezers and Refrigerator-Freezers) Regulation 1986* (N.S.W.), made pursuant to the *Electricity Act 1984* (N.S.W.), s.37(2); *Energy Labelling and Standards (Refrigerative Air Conditioners) Regulation 1987* (N.S.W.), made pursuant to the *Energy Administration Act 1987* (N.S.W.), s.38; *State Electricity Commission (Energy Efficiency Labelling) Regulations 1987* (Vic.), made pursuant to the *State Electricity Commission Act 1958* (Vic.), ss.57, 111; and the *Electrical Products Act 1988* (S.A.). This legislation is discussed in A.J. Bradbrook, "The Development of Energy Efficiency Laws for Domestic Appliances" (1990) 12 Adelaide L. Rev. 306.
72. California. California Energy Commission, *California Energy Agenda: 1989-1990 – Biennial Report* (1990) at 60.
73. Rosenfeld & Schuck, *supra*, note 70, at 4. Mortgage lenders are interested in home energy rating systems as a method of evaluating the effect of energy costs on a mortgagor's ability to repay a housing loan.

A system of mandatory energy ratings for buildings was contained in Bill S.324 introduced into the U.S. Senate in 1990. This Bill would have amended Title II of the *National Energy Conservation Policy Act*[74] by adding a new Part 6, "Residential Energy Efficiency Ratings". The Bill would have required the promulgation of guidelines for regulations to be formulated and implemented by state governments that would have required the assignment of an energy efficiency rating to residential buildings. The Bill, *inter alia*, provided for a numerical rating of the efficiency with which any residential building may be supplied with heating and cooling energy on an annual basis, provided that all residential buildings receive a rating at the time of sale, and ensured that the rating is prominently communicated to potential buyers and renters. The Bill died in committee in late 1990.

A system of mandatory energy labelling in conjunction with mandatory energy conservation building standards already exists in Florida. The Florida Thermal Efficiency Code[75] provides for the creation of an energy performance index for all new residential buildings, which must be displayed on the completed building until the time of sale. The local building official issues an energy performance index display card to be placed on the building permit by the local building inspector at the time of the final inspection by the building department. The display card is uniform statewide.

A less sophisticated system of this nature also exists in Denmark under the *Danish Building Regulations for Small Buildings 1985*. Under these regulations, the label "low energy house" is bestowed on houses whose net energy need for heating and ventilation is not more than 50 percent of the maximum energy needs of the building, calculated pursuant to a formula

---

In the U.S., the Federal Home Loan Mortgage Corporation and the Federal National Mortgage Association, which establish the lending guidelines for over 75% of the country's mortgage companies, have approved the use of energy ratings and audits in their lending practices.

74. *National Energy Conservation Policy Act*, Pub. L. 95-619, 92 Stat. 3206, 42 U.S.C.A. § 8201 *et seq.*

75. Fla. Stat. Ann. § 553.900-912 (1987).

prescribed by the legislation.[76]

Voluntary energy labelling schemes in relation to new buildings are widespread. Such schemes operate without the support of legislation. Australia has, for several years, operated the Five Star system,[77] which is based on research conducted by the Commonwealth Scientific and Industrial Research Organization (CSIRO). Building designs are examined and, if found to be satisfactory, are issued with a provisional energy efficiency five-star certificate. A full certificate may be requested on completion of the building, when a further inspection is undertaken to ensure compliance. There is no graduated scale under this system: either five stars are awarded or none. Despite limitations in its data base, the system appears to have been moderately successful in light of its small budget.

In the United Kingdom, a new National House Energy Rating Scheme has been advanced by the National Energy Foundation, a charitable trust recently established in the City of Milton Keynes.[78] The scheme measures the energy efficiency of a house. The assessment takes into account the design and form of construction of the house, the efficiency of the heating scheme, the fuel used, and the lighting system and appliances. The energy rating is on a scale of 0 to 10, with 0 being a very energy inefficient house and 10 being a very energy efficient house. The rating, together with information on the size of the house, can produce estimates of the actual energy running costs of the house. The scheme is applicable to both new and existing houses, and can act as a comparator of energy efficiency for all homes, whatever their age and condition. It is intended that the scheme will be promoted throughout the United Kingdom.[79]

---

76. Commission of the European Communities, *supra*, note 67, at 15-18.
77. *See* GMI Council of Australia Ltd., *Your Five Star Home — A Guide to Energy Efficient House Design* (Canberra, Aust.: 1987); Energy Information Centre, *5 Star Design Rating — Technical Background* (Adelaide, Aust.: 1989); J.A. Ballinger, "The 5 Star Design Rating System for Thermally Efficient, Comfortable Housing in Australia", paper presented at 3d International Conference on Housing, Climate and Building, Karlsruhe, Germany, September 1986.
78. The National Energy Foundation is an independent national organization designed to promote greater awareness of energy and its effective use.
79. National Energy Foundation, *The National Home Energy Rating* (Milton

In the United States, a voluntary Home Energy Rating Scheme, applicable to residential buildings, has recently been developed. The scheme aims to educate people on the benefits they will receive from an energy efficient house, and allows larger home loans to be obtained from lending institutions.[80] In addition, Energy Rated Homes of America, which is a non-government, industry-based organization which has gained wide community support, has adopted a voluntary system. Energy Rated Homes of America sets standards for rating the energy efficiency of homes, and trains appraisers to ascertain the rating of houses. When rating a house, the appraiser will also give advice to the owners on how most easily to improve the rating. Yard signs and brochures on energy rating are provided for participating vendors.[81]

Energy and Resource Conservation Option Points

The fourth type of energy conservation building standard for new buildings is a mandatory system of energy and resource conservation option points which are assigned to a new building for each energy conservation feature of the building.[82] Such a system exists in the subdivision and land use planning ordinances in an increasing number of U.S. municipalities. Davis, California, and Boulder, Colorado, provide representative examples. A typical points system requires all new dwelling units for which building permits are requested to score a minimum number of points from the options list that the local ordinance prescribes. Points may be awarded, for example, for buildings that are favourably oriented for solar access purposes and have roofs physically and structurally capable of supporting a minimum number of solar collectors, or for buildings that

---

Keynes, U.K.: 1989) (brochure).

80. *See* Colorado. Office of Energy Conservation, *Proposed Solicitation — Home Energy Rating System for Colorado* (Denver: 1988).

81. *See* Energy Rated Homes of America, *The Uniform System for Rating Energy Efficiency in Housing* (Little Rock, Arkansas: 1988).

82. *See* A.J. Bradbrook, "Future Directions in Solar Access Protection" (1988) 19 Env. L. 167 at 191*ff*.

contain various specified hot water saving devices, air de-stratification systems, solar devices, and insulated exterior doors, high performance glazing, ambient temperature boiler controls, and specified energy efficient appliances. Under Boulder's city code, building permit requests for all new dwelling units must score a minimum of twenty points from a list of energy and resource conservation options.[83] The list is extensive and contains a wide range of options. Illustrations are: insulation increased at least 25 percent above prescribed levels for foundations (1 point), ceilings (1 point) and walls (2 points), hot water heaters of prescribed efficiency (1 or 2 points), hot water pipe insulation (1 point), air infiltration rate below prescribed level (3 or 4 points), reduced area of non-south glazing (1 point), and insulated exterior door (1 or 2 points).

Building Vendor to Provide Insulation Information in Sales Contract

The fifth type of energy conservation building standard for new buildings is a system requiring the vendor of the building to provide information on the insulation installed in the building in the contract of sale. In the United States, this is required by the federal Trade Commission's Home Insulation Rule.[84] Analogous, more wide-ranging schemes, exist elsewhere in the United States. An interesting illustration is New York's State's *Truth in Heating Act*, which is discussed in detail below.[85] This legislation entitles all prospective purchasers and renters of houses and residential apartments within New York to receive specified information on the past heating and cooling costs of a residence they are considering buying or renting.

---

83. City of Boulder, Colo., Rev. Code § 9-3-38 (1981).

84. The Home Insulation Rule, which went into effect in 1980, requires the seller of a new home to provide information on the type, thickness, and R-value of the insulation that will be installed in each part of the house in every sales contract. Insulation contractors are required to give their customers similar information. *See* document DOE/CE-0180.

85. N.Y. Stat. § 17-103 (1983). The regulations adopted by the New York State Energy Office for implementing this Act are contained in 9 NYCRR § 7835. This legislation is discussed in more detail in Chapter 3 in the section entitled "Existing Consumer Protection Legislation".

Existing Energy Conservation Building Standards Legislation for Existing Buildings

The need to apply energy conservation measures in existing buildings is equally important as in new buildings, if energy savings are to be maximized.

Although it is architecturally straightforward to incorporate energy conservation measures into new buildings during the course of construction, in many cases it is difficult, if not impossible, to add such measures into buildings once they have been erected. The installation of some energy conservation measures into certain types of buildings (for example, wall insulation) would involve the partial demolition of the building. For this reason, the application of energy conservation building standards to existing buildings is problematic. Nevertheless, there is a considerable scope for retrofitting existing buildings with a variety of energy conservation measures. Such measures include the addition of roof insulation, increasing the thermal resistance of glazing, insulating pipes and ducts, capturing internal heat for storage or for transfer to heat-loss areas, and installing devices on the exterior of a building to shade the wall or roof where heat gain is a problem.[86]

Very few precedents exist for introducing mandatory energy conservation building standards for existing buildings. The author is unaware of any jurisdiction which requires owners of buildings to install specified energy conservation measures in existing buildings within a stipulated period, except for the requirement by a small number of U.S. states and municipalities that, in respect of existing residential buildings, energy conservation devices must be installed upon resale.

The City of Berkeley, California, applies its Residential Energy Conservation Ordinance to all residential structures or units in the city sold on or after 30 May 1987.[87] The Ordinance requires the structure or unit to satisfy specified minimum

---

86. *See* American Council for an Energy Efficient Economy, *Energy Efficiency in Buildings — Progress and Promise* (Washington, D.C.: 1986) at 34-35.
87. City of Berkeley, Calif., Ordinance No.5802 (1987) CSA A101 (Thermal Insulation, Mineral Fibre, for Buildings).

energy conservation building standards relating to the insulation of ceilings and water heaters, the sealing of furnace ducts, the installation of low-flow showerheads, the replacement of incandescent lights in multi-unit common areas, the weather-stripping of all exterior doors, and the reduction of heat loss through chimneys. The maximum required expenditure to bring a structure into compliance with the ordinance is 0.75 percent of the final sale price for a structure not containing more than two units, and 50 cents per square foot in a structure containing more than two units. In the City of Davis, California, a retrofit ordinance in effect since January 1980 requires similar conservation measures (insulation blanket for water heaters, low-flow showerheads, weather-stripping for doors, and minimum specified insulation in ceilings) to be installed in residential buildings on resale.[88] In the State of Mississippi, where an existing building is to change hands and the change will result in an increased demand for fossil fuel or electrical energy, the building is required to comply with the state Code for Energy Conservation in New Building Construction.[89] The lighting standards in the Code apply to all existing public buildings.[90]

The Preferred Model for New and Existing Buildings

Two matters are apparent from the discussion so far: first, some countries, particularly those with abundant reserves of fossil fuel, have not yet taken legislative action to improve the energy efficiency of new buildings; and secondly, there is no consistent view amongst those jurisdictions which have taken remedial action as to the correct approach to the problem. Countries which have been slow to enact energy conservation legislation for new buildings have the opportunity to learn from the experience of other jurisdictions. Countries with relevant legislation have the opportunity to "fine tune" their legislation based on the practical effects of the operation of such legislation.

---

88. *See* City of Davis, Calif., *Davis Energy Conservation Programs, Policies, and Ordinances* (Davis: 1984).
89. Miss. Rev. Stat. §§.57-39-21, 101.4(b)(3).
90. *Id.*, s.101.4(b)(4).

What should be done?

Historically, law reformers debated as to which one system — financial incentives legislation or energy conservation building standards legislation — was "preferred", that is, which system was the most energy conservation efficient. The author suggests that a choice does not have to be made — there is no reason why various types of building standards regulations cannot co-exist with various means of financial incentives. Legislation designed to promote both approaches simultaneously can be regarded as the "carrot and stick" approach to the issue. Regulations ensure that new buildings reach a minimum prescribed threshold standard of energy performance, while financial incentives can be devised so as to encourage builders and landowners to go beyond the minimum standard. Regulation without stimulation will ensure that only the barest minimum standards of energy efficiency will be attained, and additional efficiency standards will inevitably be overlooked. Stimulation without regulation runs the risk that some buildings will not have any conservation measures at all, as there will be no compulsion on builders or landowners under this system to consider energy efficiency measures. Thus, it is submitted that the "preferred" legal management model to encourage energy efficiency in new and existing buildings is a combination of the most effective financial incentives and building standards regulations. The following sections analyze the existing legislation, and all other possible options for law reform, to determine the "preferred" (the most effective) financial incentives and building standards regulations.

Preferred Financial Incentives Legislation

What form of financial incentives should be preferred to promote the use of energy conservation in new buildings: tax concessions, a system of government-backed or government loans or grants, or statutory utility energy efficiency programs? Despite the widespread use of financial incentives to promote energy conservation in the United States in the past fifteen years, six objections have been made by proponents of direct expenditure programs:

1. Direct expenditure programs are more appropriate to

meet social goals than are income tax incentives. According to the proponents of the tax expenditure analysis school of thought,[91] where government assistance is required, the first approach should be to explore the various direct expenditure alternatives. Tax incentives are argued to lead to difficulties in maintaining budgetary control, confusion in perceiving and setting national priorities, and dangers to the tax structure itself.

2. The goal of the income tax system should be to ensure that the tax legislation is as simple and as intelligible to the average layperson as possible. The creation of specific tax credits detracts from this goal.[92]
3. Income tax incentives for energy conservation measures would create administrative difficulties for the Tax Office, because the office has no experience of assessing the energy efficiency of various types of measures.[93] New laws in this area would lead to a significant increase in the number of employees in the Tax Office and would greatly add to the total cost of the tax concessions to the consolidated revenue.
4. Studies undertaken in the United States suggest that there are problems with the cost-effectiveness of a tax incentive system for energy conservation measures. According to Hyatt,[94] an immediate large-scale program of tax incentives would fail to have an impact on energy

---

91. The seminal work on tax expenditure analysis is S.S. Surrey, *Pathways to Tax Reform* (Cambridge, Mass.: Harvard University Press, 1973). *See also* Australia. House of Representatives. Standing Committee on Expenditure, *Taxation Expenditures* (Canberra: Aust. Gov. Pub. Serv., R82/340, 1982); R. Krever, "Structure and Policy of Australian Income Taxation" *in* R. Krever, ed., *Australian Taxation: Principles and Practice* (Sydney: Longman Professional, 1987) 1 at 15-25; R. Krever, "Tax Expenditures: The Other Spending Program" (1984) 10 Legal Services Bulletin 63.
92. S.V. Hyatt, "Thermal Efficiency and Taxes: The Residential Energy Conservation Tax Credit" (1977) 14 Harvard J. Legis. 281 at 318.
93. *Id.*, at 319.
94. *Id.*, at 324.

conservation commensurate with the size of the program and the magnitude of government revenues necessary for its implementation.

5. A system of income tax incentives to promote energy conservation may produce an inefficient use of resources. McDonald writes:[95]

> The [tax] credit lowers the effective private cost of investments designed to conserve energy or increase use of renewable sources of energy such as solar energy. Of course, it does not lower the social costs of such investments .... [The credit] will induce too much investment in conservation and substitution, as the investment is pushed beyond the point where marginal social benefit equals marginal social cost. The appropriate amount of investment would occur if the prices of conventional forms of energy were free to reflect their relative scarcity. As conventional fuels rise in relative price, as they are almost sure to do for the foreseeable future, solar energy and the like will be used in their own good time.

6. Tax incentives to promote energy conservation may induce inflation in the energy conservation materials industry and erode the purchasing power of subsidy dollars.[96] This will arguably result from demand-pull inflation caused by a sudden increase in the purchasing of these materials following the imposition of the tax incentive.

The merits of these six objections are dubious. For example, the goal of a simple tax system was lost many years ago, as the present income tax laws in all western industrialized nations are extremely complex and contain numerous technical exceptions. A further concession granted in favour of energy conservation measures would make little difference to the conceptual nature of the income tax legislation. While financial incentives through income tax to promote energy conservation could well create administrative difficulties for the Tax Office, there are many other situations where the Tax Office has had to expand the scope of its traditional work in recent years as a result of legislative changes. A useful illustration is the new Australian

95. McDonald, *supra*, note 4, at 866.
96. Hyatt, *supra*, note 92, at 297.

child support scheme, which provides for the collection of maintenance by the use of an income tax surcharge levied against the non-custodial parent.[97] The argument that a system of income tax incentives in building energy conservation may produce an inefficient use of resources may be countered, as McDonald admits,[98] by the fact that the industries supplying energy conservation equipment and installation are not mature and cannot set a price on the basis of high volume. The effect of this is that some initial artificial inducement may result in a more efficient use of resources in the long term. Finally, in relation to tax expenditure analysis,[99] even the anti-tax incentive proponents concede that a program of government assistance that is broadly based, relatively simple, and properly structured can be appropriately joined to the tax system.[100] These criteria are the basis of the four policy issues which must be addressed when proposing new income tax incentives.[101]

In addition to these arguments, there are two reasons which justify financial incentives through income tax to promote energy conservation:

1. As a matter of basic justice and economic sense, all forms of energy should compete for use in the marketplace on an equal footing. In most countries, the existing laws give numerous financial incentives to traditional fossil fuels.[102] Tax incentives for energy conservation measures are necessary to allow related industries to compete.[103]

---

97. *Child Support (Assessment) Act 1989* (Cth.).
98. McDonald, *supra*, note 4, at 866.
99. See *supra*, note 91.
100. Surrey, *supra*, note 91, at 149.
101. See the section of this chapter entitled "Criteria to Formulate Tax Incentives".
102. In Australia, for example, *see Income Tax Assessment Act 1936* (Cth.), Part III, Div. 10AA (Prospecting and Mining for Petroleum). S. Boyle, *in* "More Work for Less Energy" (5 August 1989) *New Scientist* 19 at 22, states that in the U.S. in 1984, energy subsidies totalled $44 billion, more than $500.00 for every household in the country. Approximately 95% of them go to conventional technologies, such as those exploiting nuclear, coal, and gas power; less than 2% assist energy efficiency.
103. In the context of solar energy, this point was made forcefully by the

2. A "catch-22" situation exists: if commercialization of renewable energy devices and energy conservation measures is to proceed, the high cost must be reduced, but if the necessary cost reduction is to be achieved, these devises and measures require significant market penetration and increased production. This problem has also been noted with concern by the California Energy Commission.[104]

The majority of the arguments listed above in favour of tax incentives can also be used to support the introduction of a system of government or government-backed loans or grants. However, as the installation of energy conservation measures in private buildings may enhance the value of the property, it is submitted that it is more appropriate to encourage and reward individuals to invest their own money in energy conservation measures rather than to rely on government or government-backed grants or loans. If a system of government or government-backed loans or grants is adopted in the context of residential buildings, it would seem fair and reasonable to make an exception in favour of low-income families. The exception could take the form of a grant or a low-interest loan made for the installation of specified energy conservation measures. Such a grant or loan could be made directly by the state government or through the relevant electricity or gas utility.

In summary, there are strong arguments favouring the use of both government or government-backed loans or grants and

---

California Solar Energy Industries Association (CAL SEIA): *see* California. California Energy Commission, *Solar and Wind Technology: Tax Incentive Impact Analysis* (Sacramento, Ca.: Report P500-86-010A, 1986) App.A. *See also* J.V. Lamar, "Gone With the Wind", *Time*, 20 January 1986, where it is argued that support of renewable energy devices cost the U.S. federal government $550 million in lost taxes in 1984, while lost taxes and direct subsidies for the nuclear power industry alone cost $15 billion; and Canada. Dept. of Energy, Mines and Resources, *Energy in Canada — A Background Paper* (Ottawa: 1987) at 96-105.

104. California. California Energy Commission, *Solar and Wind Technology: Tax Incentive Impact Analysis* (Sacramento, Ca.: Report P500-86-010A, 1986) at 2-11.

tax incentives.

One question remains for consideration: should a system of tax incentives and/or government or government-backed loans or grants be coupled with statutory utility programs to improve energy efficiency in the community. This issue has not been discussed by other legal commentators. The various types of statutory utility programs in place in the United States appear to have been successful in reducing electricity demand. Such programs have certainly been partly responsible for reducing the need for new power stations.[105] Although government interference with the operation of private electric utilities is always controversial, the utilities are already heavily regulated by statute in light of the public interest inherent in the provision of energy supplies.[106] The compulsory provision of statutory energy conservation programs would involve quite minor additions to the utilities' existing duties and to the legislation regulating electric utilities.

It is submitted that the public interest inherent in fuel

---

105. See, for example, C. Schroeder and A. Miller, "The Validity of Utility Conservation Programs According to Generally Accepted Regulatory Principles" (1982) 3 Solar Law Reporter 967.

106. In the United States, for example, regulation is exercized federally by the Federal Energy Regulatory Commission and at state level by a Public Utilities Commission or Public Services Commission. Each commission is established and operates pursuant to legislation. For a discussion of the U.S. regulatory system, *see* J.P. Tomain and J.S. Hickey, *Energy Law and Policy* (Cincinnati: Anderson Publishing Co., 1989) c.5 & 10; D.N. Zillman and L.H. Latman, *Energy Law* (Mineola, N.Y.,: Foundation Press, 1983), c.2. A similar system of provincial regulation of electric utilities exists in Canada. See, for example, *Ontario Energy Board Act*, R.S.O. 1980, c.332; *Public Utilities Board Act*, R.S.M. 1987, c.P280; *Public Utilities Review Commission Act*, S.S. 1982, c.P-45.1; *Public Utilities Bard Act*, R.S.A. 1980, c.P-37. As in the U.S., there is federal control pursuant to the *National Energy Board Act*, R.S.C. 1970, c.N-6, although for constitutional reasons the Board has little effective control over electricity sales within Canada. The system in the United Kingdom consists primarily of control by competition rather than regulation: *see Electricity Act 1989*, and S.K. Bailey and R.H. Tudway, *Electricity Law and Practice* (London: Sweet and Maxwell, 1992).

conservation can certainly justify the mandatory imposition of suitable utility programs to foster energy conservation and the use of renewable energy resources. Such programs would be administratively straightforward to implement, would be relatively cheap to operate, and would enhance the public perception of utilities as responsible energy managers, rather than simply as energy vendors. Thus, it is submitted that the introduction of statutory utility programs to promote energy conservation should be regarded as a preferred option, subject to the qualification that, in jurisdictions where electric utilities are in private ownership, a system of financial compensation should be established to ensure that the utility is reimbursed both for the costs associated with implementing and operating energy conservation programs and for the loss of profitability caused by declining electricity sales.

As stated earlier, although there are strong arguments favouring the use of both government or government-backed loans or grants and tax incentives, the preference of the author is a system of tax incentives, to encourage and reward individuals who invest their own money to install energy conservation measures. While a system of loans or grants could be implemented without the need for legislative change,[107] law reform will be required to establish new financial incentives through taxation.

Criteria to Formulate Tax Incentives

If tax incentives are adopted, the following four policy issues should be addressed: determination of the relevant criteria for the formulation of the income tax incentives, whether building performance criteria should be required as a condition precedent to any income tax incentives, whether the tax incentives should constitute a deduction or a credit/rebate, and the length of time new tax incentives for building energy

---

107. Legislation may be required to clarify the tax status of any loans or grants received. For example, s.12(1)(u) of the Canadian *Income Tax Act*, S.C. 1970-71-72, c.63, requires taxpayers to include in their income the amount of any grant received under the (now-discontinued) Canadian Home Insulation Program for home insulation.

conservation should be maintained. Each of these policies will be examined in turn.

The first policy issue to be addressed concerning the development of new tax incentives is the determination of the relevant criteria. The California Energy Commission specified nine relevant criteria for the formulation of income tax incentives for renewable energy devices and energy conservation measures:

1. The tax incentives must be effective for the time period necessary for commercialization.
2. The incentives must be adequate in level to stimulate commercial investment in the most viable energy conservation technology options.
3. The incentives must provide for smooth, predicable phasing-out.
4. The incentives must be selective in order to achieve high performance and long operating life.
5. The incentives must stimulate, rather than inhibit, technical progress and innovation.
6. The potential for abuse must be minimised.
7. The incentives must be simple to implement and administer.
8. The incentives must address the financial imbalance in incentives between conventional technology and energy conservation technology.
9. The incentives must be market-oriented rather than technology-oriented.[108]

The second policy issue to be addressed concerning the development of new tax incentives is whether building performance criteria should be required as a condition precedent to any income tax incentives. The federal, and the vast majority of the state, tax incentives in the United States for renewable energy devices and energy conservation measures apply regardless of building performance. The experience in that country suggests that this is a mistake, and that future tax

---

108. California. California Energy Commission, *Solar and Wind Technology: Tax Incentive Impact Analysis* (Sacramento, Ca.: Report P500-86-010A, 1986) at 7-2.

concessions should be restricted to devices which meet specified government-approved criteria.[109] For example, in the wind-energy industry in the United States, the existence of generous tax incentives during the 1970s encouraged investors and entrepreneurs to establish wind farms largely as a tax shelter rather than as a serious industrial enterprise. Many of the wind farms[110] were grossly inefficient and were referred to colloquially as "tax farms".[111] The Michigan Office of Energy Programs in the state Department of Commerce has reported that tax credits to promote renewable energy or energy conservation measures without efficiency or building performance standards have proven to be ill-advised.[112]

The third policy issue to be addressed concerning the development of new tax incentives is whether the tax incentives should constitute a deduction or a credit/rebate. This issue raises the question of tax equity. As the goal of any income tax

---

109. Note that in 1989, the California State legislature passed S.B. 227 (Stats. 1989, c.1291), which establishes a state tax credit for commercial solar energy systems of 30 watts or more. In S.B. 227, the legislature directed the California Energy Commission to establish minimum eligibility requirements for safety, durability, reliability, market readiness, efficiency, and cost-effectiveness in terms of dollars per kilowatt. The requirements are still under consideration by the Commission: California. California Energy Commission (Docket No.89-STC-1, 19 December 1990). This legislation may well serve as a model for all future U.S. laws relating to energy conservation measures and renewable energy devices.
110. A "wind farm" is a collection of wind generators located on windy tracts of land designed to generate electricity for the local electric power company. Many wind farms exist in the United States, particularly in California and Hawaii. For a discussion of wind farms, *see* M. Evans, "Windfarming in America", (1982) 24 Wind Power Digest 6; R. Odland, "Planning the Wind Farm: State and Local Concerns" *in Proceedings of the Fifth Biennial Wind Energy Conference and Workshop* (Washington, D.C: U.S. Dept. of Commerce, Report SERI/CP-635-1340, 1981) v.3.
111. *See* J.V. Lamar, "Gone With the Wind", *Time*, 20 January 1986, at 22.
112. Letter from T. Stanton, Office of Energy Programs, Michigan Dept. of Commerce, to A.J. Bradbrook, 23 January 1987. *See also* Michigan. Dept. of Commerce, *Michigan Tax Incentives for Residential Renewable Energy Systems* (1986).

concession should be to promote the use of energy conservation by all persons, regardless of their level of income, care should be taken to ensure that the tax incentive is equitable. In many countries, the majority of income tax concessions constitute deductions, rather than credits or rebates. In the present context, however, deductions would constitute the least equitable means of incentive. This arises because the taxpayer's marginal tax rate determines the amount of tax dollars saved, with the value of the deduction increasing with the amount of the taxpayer's assessable income. Thus, the rich obtain greater value from the deduction than do the poor.[113] A tax credit or rebate has an advantage over deductions in that both constitute an absolute reduction in tax liability which does not vary in value according to the level of the taxpayer's income. This applies regardless of whether a full credit is given for the value of the energy conservation measure or whether the rebate is limited to the prevailing base marginal tax rate. Despite this, tax inequities still exist. Wealthy homeowners would tend to receive more tax benefit as they have more capacity to invest in energy conservation measures and will usually spend more in this respect because, in general, they live in larger homes than poorer people.[114] As studies in the United States have shown,[115] greater attention should be given to the poor and the lower middle class, as they use far more units of energy per square metre of living space than do the wealthy. A further problem is that a tax credit or rebate will be of no benefit at all to a person whose taxable income is below the income tax threshold.[116] Whichever system of tax incentives for energy conservation measures is adopted, persons on a lower income will be disadvantaged relative to persons on a higher income, although considerations of tax equity would tend to favour a system of tax credits or rebates rather than deductions.

---

113. Hyatt, *supra*, note 92, at 302-303.
114. Hyatt, *supra*, note 92, at 292; Ferrey, *supra*, note 15, at 487-488.
115. Ford Foundation, Energy Policy Project, *A Time to Choose: America's Energy Future* (New York: 1974) at 36.
116. For example, in Australia the income tax threshold is $5,400 for the 1991-1992 year of taxation.

The fourth policy issue to be addressed concerning the development of new tax incentives is the length of time new tax incentives for building energy conservation should be maintained. It is submitted that tax incentives should not remain indefinitely, but should only be maintained until energy conservation measures are able to compete in the marketplace on an equal basis with conventional energy technologies. Incentives can be justified on the basis that energy conservation measures are relatively new and untried, that demonstration effects would hasten sound development, and that industries supplying equipment and installation are not mature and do not have the price advantage of high turnover.[117] As recommended by the California Energy Commission,[118] energy conservation costs decline through increased production and technological subsidies can gradually be withdrawn and phased out.

As it is impossible to predict in advance when energy conservation measures will become cost competitive with conventional energy technologies, it is submitted that any new legislation enacting tax incentives should not specify an expiry date. Instead, the progress of the industries should be continually monitored by the government, which at the appropriate time would introduce amending legislation declaring the incentives inapplicable on and after a designated date.

Preferred Energy Conservation Building Standards Legislation for New Buildings

Energy conservation building standards for new buildings are of five types: building performance regulations, prescriptive regulations for building materials, a system of energy rating or labelling of buildings, a conservation options points system, or compulsory disclosure of existing conservation measures at the time when a building is first sold.

When determining which of these types is to be preferred, it is submitted that the decision should be based on the following

---

117. McDonald, *supra*, note 4, at 866.

118. California. California Energy Commission, *Solar and Wind Technology: Tax Incentive Impact Analysis* (Sacramento, Ca.: Report P500-86-010A, 1986) at 2-11.

four principles:

1. Any new system of regulations should be compulsory.
2. As any regulations will operate throughout the jurisdiction, such regulations must be equally relevant and appropriate to all parts of each jurisdiction.
3. The view of the building industry that it is already over-regulated should be taken into consideration.
4. A flexible, as opposed to a rigid, system of regulations is likely to be more appealing to the building industry and therefore more politically acceptable.

The application of these four principles leads to the conclusion that prescriptive regulations for building materials, energy rating or labelling of buildings, and the system of compulsory disclosure of existing conservation measures at the time a building is first sold should not be pursued. In relation to prescriptive regulations for building materials, problems arise because of climatic differences within various parts of each jurisdiction. What is regarded as an appropriate level of ceiling insulation in one location may well be excessive or inadequate in another location. In Alberta, for example, more insulation would be required in Fort McMurray than in Lethbridge. Far greater regional differences exist in some other jurisdictions. This problem could be overcome, at least partially, by introducing into the legislation the concept of degree heating days and imposing more onerous conservation measures for areas with higher heating-day values. However, this could only be achieved at the cost of introducing enormous complexity into the building laws which are already notorious for their complexity.

Although a mandatory rating or labelling system is not recommended, *were* it to be introduced, legislation would need to specify the circumstances in which a rating or label would need to be obtained, the method for its calculation and the manner in which, if at all, it must be displayed or communicated, together with appropriate penalties. Responsibility for administering the scheme would need to be given to a government department, existing or specially created, and the powers and duties of inspectors must be determined.

If a rating or labelling system were to be voluntary, legislation would not normally be required. However, if such a system were to function as a performance-based alternative to a set of energy-related specifications in the relevant Building Code, the details of the labelling scheme would need to be set out in an enactment or regulations, preferably the Building Code itself.

However, energy rating or labelling of buildings and compulsory disclosure of existing conservation measures at the time a building is first sold both suffer from the problem that neither system requires any energy conservation improvements to be made. At best, all either system does is to put some indirect pressure on builders to give consideration to energy conservation, which pressure can easily be ignored or discounted. Experience with the system of energy labelling for domestic appliances in effect in some countries has shown that many consumers have tended to ignore the energy consumption figures displayed on the label and have been more influenced in their choice of purchase by other factors.[119] Energy rating or labelling of buildings is likely to suffer a similar fate.

The author submits that either building performance regulations or a conservation options points system would be appropriate for adoption in any jurisdiction. Both options appear to satisfy the four principles discussed above.[120] As both options are performance-based, they would apply equally to all buildings regardless of location and climatic factors. For example, if the regulations specify minimum standards of airtightness, more insulation will be required in the cooler areas of any jurisdiction than in the warmer areas. In addition, either system, although compulsory, would give considerable flexibility to builders and architects and is likely to be regarded by them as more acceptable than the alternative types of regulation.

A further alternative would be to devise and legislate a system which combines building performance regulations and an energy conservation options points system. Under such a system, building performance regulations could be established in respect

---

119. Information supplied by Dr. A. Pears, Energy Planner, Dept. of Industry, Melbourne, Victoria, Australia.

120. See the section of this chapter entitled "Preferred Energy Conservation Building Standards Legislation for New Buildings".

of a variety of individual measures, rather than the building as a whole, and these measures could each be assigned a number of conservation points based on their relative energy saving potential. As under the system in the City of Boulder, Colorado,[121] each building would be required to attain a minimum specified number of conservation points to satisfy the building regulations. A system of this nature would also satisfy all of the four principles identified above.[122]

Summary

The author suggests that it is not necessary to choose between financial incentives or mandatory energy conservation building standards legislation to promote energy efficiency in new buildings, as the various types of regulation can co-exist with various means of financial incentives. Thus, he submits that the "preferred" legal management model to encourage energy efficiency in new buildings is a combination of the most effective financial incentives and regulations. The existing financial incentives (tax concessions, government-backed or government grants and loans, and statutory utility energy efficiency programs) are all found to be preferred financial incentives. Of the five existing types of energy conservation building standards, three (prescriptive regulations for building materials, a system of energy rating or labelling of buildings, and compulsory disclosure of existing conservation measures at the time when a building is first sold) were found to be not particularly satisfactory, while two (building performance regulations and an energy and resource conservation options points system) were found to be "preferred". The author proposes a further alternative which would combine building performance regulations and an energy and resource conservation options points system.

The existing financial incentives (tax concessions, government-backed or government grants and loans, and statutory utility energy efficiency programs), as discussed in the

---

121. See *supra*, note 83 and accompanying text.
122. See the section of this chapter entitled "Preferred Energy Conservation Building Standards Legislation for New Buildings".

context of new buildings, apply with equal force to existing buildings. In addition, the majority of common law jurisdictions have adopted specific funded programs to improve energy conservation in existing buildings. There are two reasons why primary reliance must be placed on financial incentives to encourage the installation of energy conservation measures in existing buildings on a voluntary basis. One, the application of energy conservation building standards to existing buildings is problematic, because of the difficulty (and in some cases, the impossibility) of adding such measures into buildings once they have been erected. A second reason is that it is regarded by many as politically unacceptable to apply retrospectively new regulations to existing buildings. While retrospective legislation might be justifiable in extreme cases where human life is at risk, such is not usually the case. Of the five existing energy conservation building standards, only the system of compulsory disclosure of existing conservation measures is preferred for existing buildings, in that it is politically acceptable. Such a disclosure could be required by legislation whenever a building is sold or leased, and would put indirect pressure on existing owners to upgrade the conservation measures prior to disposing of their interest in the premises.

The author suggests that the preferred legal management model for energy conservation in new and existing buildings is some combination of the most effective financial incentives and building regulations. Governments may, however, disagree with the author. The reality is, that, ultimately, the preferred governmental choice regarding financial incentives and building standards, for both new and existing buildings, is a political rather than a legal issue, and will be determined by governments in consultation with representatives of architects and the building industry, and with their economic advisers.[123]

---

123. In the United States, a number of studies designed to assess the extent to which tax incentives stimulate conservation have produced inconclusive results. *See* United States. General Accounting Office, *Appliance Efficiency Standards: Issues Needing Resolution by DOE* (Washington, D.C.: Report GAO/EMD-82-78, 1982); E. Hirst, R. Goeltz & H. Manning, "Analysis of Household Retrofit Expenditures" (1983) 7

3. ENERGY CONSERVATION LAW FOR RENTAL BUILDINGS

This chapter initially sets out, factually, the "existing" legislation used by various governments to promote energy efficiency in rental buildings. The existing legislation, and all other possible options for law reform, is then analyzed to determine its effectiveness. Some law reforms are suggested. A "preferred" legal management model to encourage energy efficiency in rental buildings is proposed. The chapter concludes with a summary. But first, it is necessary to understand the background policy issue that sets rental buildings apart from new and existing buildings as it relates to energy conservation measures.

Background Policy Issue

The policy issue to be determined is this: which party (landlord or tenant) has the duty to install and maintain energy conservation measures sufficient for the rental premises to meet a designated standard of energy efficiency? The nub of the issue is that neither commercial or residential tenants nor landlords have any incentive to install energy conservation measures in rental buildings. Tenants and landlords have different reasons for their reluctance to invest in energy conservation.

There are four reasons why tenants are unwilling to invest in building energy conservation: they are reluctant to make capital investments on the landlord's property, the high turnover rate of tenants means they do not live in one place long enough to be concerned about energy conservation, poverty among residential tenants makes purchase of any conservation improvements difficult, and tenants appear to feel that energy conservation is the landlord's responsibility.

Energy Systems and Policy 303; United States. Dept. of Energy. Energy Information Administration, *An Economic Evaluation of Energy Conservation and Renewable Energy Tax Credits* (Washington, D.C.: 1985).

From a residential or commercial tenant's perspective, the first problem is that, as tenants do not own the premises, they are extremely reluctant to make capital investments on the landlord's property by installing energy conservation measures.[1] Any such measures installed by a tenant in the rented premises will become fixtures under traditional common law rules and legal title will vest in the landlord.[2] The landlord is under no obligation to compensate the tenant for the value of the improvements. Although tenants are given certain rights at common law to remove fixtures prior to the termination of the tenancy agreement, such right of removal is limited to trade, ornamental, and domestic fixtures.[3] Further, the cost of removing the fixtures, including the cost of restoring the area where the fixture was removed from, make it unlikely that tenants would remove fixtures. In light of the existing case law as to what falls within the scope of these exemptions, it appears most unlikely that a tenant would be legally permitted to remove any energy conservation measures.

In the case of residential tenancies, a second problem is the high turnover rate of tenants. In the United States, census data indicates that the median length of occupancy of private rented housing is between 2 and 5 years, with the majority at the shorter end of the range. This compares with an average tenure for owner-occupied housing at between 6 and 10 years.[4] The high rate of mobility makes tenants reluctant to invest money to

---

1. R.H. Counihan & D. Nemtzow, "Energy Conservation and the Rental Housing Market" (1981) 2 Solar Law Reporter 1103 at 1105.
2. For a discussion of the common law rules relating to fixtures, *see* R.E. Megarry & H.W.R. Wade, *The Law of Real Property*, 5th ed. (London: Stevens, 1984) at 730-738; *Cheshire and Burn's Modern Law of Real Property*, 13th ed. (London: Butterworths, 1982) at 136-141.
3. See, for example, *Smith* v. *City Petroleum Co.*, [1940] 1 All E.R. 260; *Spyer* v. *Phillipson*, [1931] 2 Ch.Div. (Eng.) 183; *Concepts Property Ltd.* v. *McKay*, [1984] 1 N.Z.L.R. 560 (H.C.); *New Zealand Govt. Corp.* v. *HM & S Ltd.*, [1982] Q.B. 1161; *D'Arcy* v. *Burelli Investments Pty Ltd.* (1987), 8 N.S.W.L.R. 317 (S.C.).
4. California. California Energy Commission, *Energy Conservation in California Multifamily Rental Housing* (Sacramento, Ca.: Report P400-85-009, 1985) at 15.

install energy conservation measures in rented premises.

Census statistics show that tenants have below average incomes. A third problem, then, is that poverty among residential tenants makes purchase of any building energy conservation improvements unlikely. This is particularly unfortunate as, although low income people use less energy per capita than others, they may spend as much as one third of their disposable income to meet their energy requirements.[5]

The fourth problem is that tenants appear to feel that energy conservation is the landlord's responsibility. This was confirmed in the United States by a 1982 study of tenants of the Palo Alto Comprehensive Multi-Family Conservation Project residential premises in Palo Alto, California. This project found that 80 percent of tenants surveyed believe that energy conservation is not their concern, but the concern of the landlord. The project further reported that even if some tenants felt responsible for energy conservation, 37 percent lacked information on what to do or buy, and 25 percent lacked information as to where to buy the necessary materials.[6] Although there has been no similar study undertaken elsewhere, there seems to be no reason to suspect that the situation would be any different in any other western industrialized country.

There are two reasons why landlords are unwilling to invest in building energy conservation. The first reason is that in most countries tenants are responsible for paying the gas and electric utility accounts. An exception exists in North America, where a significant percentage of multi-unit dwellings are master metered with the landlord responsible directly to the utilities for the payment of the accounts.[7] This does not significantly affect the

---

5. Note, "Minorities and Rental Housing Solutions" (1981) 3 Solar Law Reporter 181 at 181.
6. City of Palo Alto, Calif., Utilities Dept., *Marketing Conservation for Multi-Family Housing* (Palo Alto: 1982). Discussed *in* California. California Energy Commission, *Energy Conservation in California Multifamily Rental Housing* (Sacramento, Ca.: Report P400-85-009, 1985) at 46-47.
7. For a discussion of master metering, *see* Counihan and Nemtzow, *supra*, note 1, at 1112-1120; A. Levine & J. Raab, *Solar Energy, Conservation, and Rental Housing* (Washington, D.C.: U.S. Dept. of Commerce,

situation, however, as landlords can pass on the additional energy charges to the tenants by way of increased rent. The second reason, from the landlord's perspective, is the fact that any investment in energy conservation measures is unlikely to increase the rental or resale value of the premises.[8] Studies in the United States have shown that the working capital of landlords of rental buildings is often limited, and that landlords seek to maximize the return on their investments.[9] Both these factors militate against landlords installing energy conservation measures in their rental buildings.

The discussion above shows that there is a strong need for energy conservation legislation specific to rental buildings. This will only be achieved by legislative reform. Before considering the various options for law reform measures, however, it is necessary to consider the extent to which the existing law encourages the development of energy conservation for rental buildings.

Existing Legislation

Few specific statutory measures have been designed to promote energy conservation in existing rental buildings, nevertheless, relevant measures have been enacted in some jurisdictions. The existing legislation which appears to encourage the installation of energy conservation measures into rental homes includes financial incentives, mandatory building standards, and consumer protection law. The laws in Canada, Australia, and the United States will be used as illustrations.

Existing Financial Incentives Legislation

The only existing financial incentives legislation relevant to

---

Report SERI/RR-74-901, 1980) at 30-32. According to Counihan and Nemtzow, in the United States, as at 1981, tenants in 59 percent of rented housing are responsible for heating and hot water bills, while owners of 41 percent of rental units pay for heating and hot water.

8. California. California Energy Commission, *Energy Conservation in Rental Housing Conference Proceedings* (Sacramento, Ca.: Report P400-85-013, 1985) at 3.

9. Counihan & Nemtzow, *supra*, note 1, at 1105.

the development and promotion of energy conservation measures for rental buildings is income tax legislation, which in certain very limited circumstances may provide tax relief for investments made by landlords in purchasing and installing energy conservation measures in their rental buildings. However, in most situations, tax relief by way of a deduction from assessable income of the cost of purchasing and installing energy conservation measures in rental buildings will not be available because of the wording of the legislation. The Canadian *Income Tax Act*, s.18 reads:

> (1) In computing the income of a taxpayer from a business or property no deduction shall be made in respect of:
> (a) an outlay or expense except to the extent that it was made or incurred by the taxpayer for the purpose of gaining or producing income from the business or property;
> (b) an outlay, loss or replacement of capital, a payment on account of capital or an allowance in respect of depreciation, obsolescence or depletion except as expressly permitted by this Part.[10]

In the present context, although the expense incurred by a landlord in purchasing and installing energy conservation measures in rental buildings would be incurred in "gaining or producing income from the business or property", the deduction would be denied because energy conservation measures would undoubtedly be regarded as items of a capital nature.[11]

Tax relief for landlords in the situation under discussion arises in relation to the availability of a deduction for the cost of repairs. In Canada, this deduction is available under common law principles.[12] In Australia, it is codified in s.53(1) of the *Income*

---

10. *Cf.*, *Income Tax Assessment Act 1936* (Aust.), s.51(1):

> All losses and outgoings to the extent to which they are incurred in gaining or producing the assessable income, or are necessarily incurred in carrying on a business for the purpose of gaining or producing such income, shall be allowable deductions except to the extent to which they are losses or outgoings of capital, or of a capital, private or domestic nature, or are incurred in relation to the gaining or production of exempt income.

11. See, for example, *Siscoe Gold Mines Ltd.* v. *M.N.R.*, [1945] Ex. C.R. 257; 2 D.T.C. 749; *Highwood–Sarcee Oils Ltd.* v. *M.N.R.*, [1944] S.C.R. 92; *Graham and Vick Ltd.* v. *M.N.R. (T.A.B.)* (1950), 50 D.T.C. 43; *Supertest Petroleum Corp.* v. *M.N.R. (T.A.B.)* (1964), 64 D.T.C. 244; *New Anchor Hotel Ltd.* v. *M.N.R. (T.R.B.)* (1976), 76 D.T.C. 1325.
12. See, for example, *Rhodesia Railways* v. *Collector of Income Tax*,

Tax Assessment Act 1936, which reads:

> Expenditure incurred by the taxpayer in the year of income for repairs, not being expenditure of a capital nature, to any premises, or part of premises, plant, machinery, implements, utensils, rolling stock, or articles held, occupied or used by him for the purpose of producing assessable income, or in carrying on a business for that purpose, shall be an allowable deduction.

While the installation of energy conservation measures in rental buildings will not normally amount to "repairs", it may be so if the repair work involves the replacement of an energy inefficient item with an energy conservation measure. Possible illustrations of this include the installation of low-flow showerheads instead of ordinary showerheads as part of the repair work done to a bathroom. Similarly, a worn-out water heater may be replaced by one with a water-heater blanket attached. Repair work to the outside walls or doors of a house may arguably justify the inclusion of weather-stripping as an integral part of the work.

The issue as to whether in such situations the expenditure made in respect of energy conservation measures is tax deductible will depend on whether the expenditure is of a capital nature, and thus excluded by the wording of the legislation. The effect of this is that if the replacement which is alleged to constitute "repairs" produces an improvement to the property, the expenditure is not deductible.

The meaning of "improvement" in this context was explained by Denning LJ in *Morcom* v. *Campbell-Johnson* as follows:

> If the work which is done is the provision of something new for the benefit of the occupier, that is, properly speaking, an improvement; but if it is only the replacement of something already there which has become dilapidated or worn out, then albeit that it is a replacement by its more modern equivalent, it comes within the category of repairs and not improvements.[13]

In this case, the Court of Appeal held that the expenditure

Bechuanaland, [1933] A.C. 368 (P.C.); *Graham and Vick Ltd.* v. *M.N.R. (T.A.B.)* (1950), 50 D.T.C. 43; *W.E. Coutts Co.* v. *M.N.R. (T.A.B.)* (1956), 56 D.T.C. 26.

13. *Morcom* v. *Campbell-Johnson*, [1955] 3 All E.R. 264 at 266. *See also FCT* v. *Western Suburbs Cinemas Ltd.* (1952), 86 C.L.R. 102 (H.C.A.); *Wates* v. *Roland*, [1952] 2 Q.B. 12.

incurred by the taxpayer in substituting six worn-out 80 gallon water tanks by one 480 gallon tank, and in substituting a worn-out two pipe system for removing waste from a toilet and wash-hand basin by a cheaper and modern one-pipe system could be deducted under the English legislation relating to deductions for repairs even though the expenditure amounted to a modernization of the taxpayer's property. This case can be usefully contrasted with *FCT* v. *Western Suburbs Cinemas Ltd.*,[14] where Kitto J held that expenditure incurred in replacing a theatre ceiling with a new and improved ceiling amounted to expenditure of a capital nature and therefore could not be deducted. The case is distinguishable from *Morcom* v. *Campbell-Johnson* in that Kitto J concluded on the facts that the new ceiling went beyond the need for restoration.

It is thus a question of fact in each case whether the installation of energy conservation measures as part of repair work constitutes an improvement. It is submitted that an energy conservation measure does not amount to an improvement merely because the replacement item is more energy efficient than its predecessor. The crucial question is whether the replacement item is merely the modern equivalent of the replaced item or whether the replacement goes beyond the need for restoration. In the former case, the expenditure incurred by the taxpayer will be allowable as a deduction, but in the latter case it will be disallowed as being expenditure of a capital nature. Nevertheless, however generously the courts view the application of the deduction for repairs in the context of the issue under discussion, it is clear that it is only of marginal relevance to the overall purpose of maximizing the use of energy conservation measures in rental buildings.

Existing Energy Conservation Building Standards Legislation for Rental Buildings

It must be remembered that a number of jurisdictions have mandatory building standards which, while not specifically directed to rental buildings, nevertheless contain minimum

---

14. *FCT* v. *Western Suburbs Cinemas Ltd.* (1952), 86 C.L.R. 102 (H.C.A.).

prescribed construction standards for building energy efficiency. These standards incidentally benefit all newly constructed rental buildings within the jurisdiction.[15]

Significantly, the issue of special conservation rules for rental buildings is not addressed by any Canadian or Australian building regulations or by-laws. The only mandatory energy conservation building standards for rental buildings are found in state and municipal legislation in the United States. Wisconsin legislation requires the state Department of Housing to adopt rules which establish a code of minimum energy efficiency standards for rental units.[16] The rules require the installation of specified energy conservation measures.[17] The present value

---

15. Calif. Code of Regulations, Title 24, Part 2, c.2-53, "Energy Conservation Standards", contains minimum standards of energy efficiency for all types of new buildings, including all residential and commercial structures. In addition, in some jurisdictions, certain municipalities have enacted local building codes which require all new residences, including rental houses and apartments, to comply with minimum prescribed standards of energy efficiency in their construction.
16. Wis. Stat. Ann. § 101.122. The rules are contained in chapters ILHR 67 and 68, which became effective on 1 January 1985. The legislation and rules are referred to as the "Rental Weatherization Program".
17. The following minimum requirements must be adhered to before a certificate of compliance may be issued:
    1. All "accessible" areas must be insulated to the levels shown in Chapter 67.05 of the code.
    2. Windows must be double-glazed or equipped with storm windows.
    3. Exterior doors and windows must be weatherstripped where cracks exceed 1/10 inch.
    4. All exterior joints on the first three floors must be caulked.
    5. All exterior, inwardly-swinging, uninsulated doors must be equipped with storm doors or air lock vestibules.
    6. Moisture control ventilation must be installed in attics. The free area of the vents must be a minimum of 1/300 of the floor area. Crawlspace floors must have a vapour barrier installed.
    7. Water heaters and heating equipment must have been inspected within the last 6 months. Water heaters in uninsulated basements must be retrofitted with an insulated jacket of R-5 or better. The first five feet from the water heater of both hot and cold domestic water pipes must be insulated to an R-2 in non-circulating system and total length supply and return in

benefits of each energy measure, in terms of saved energy over a five-year period after installation, must be more than the total present value cost of installing the measures.[18] The various standards are enforced by inspectors, appointed under the legislation for the purpose of issuing certificates of compliance with the state legislation. Pursuant to the Act, no landowner may transfer a rental unit unless, within the previous five years, an inspector has inspected the unit and has issued a certificate stating that the unit satisfies the prescribed standards.[19] Exceptions apply where a rental unit is scheduled for demolition within two years, or where the transferee of a rental unit presents a stipulation signed by the transferee and by the relevant municipality stating that the owner of the rental unit will bring the rental unit into compliance with the prescribed standards within one year of the date of transfer. Historic buildings are also exempted.[20] Similar, although not identical, legislation exists in Minnesota.[21]

Chapter 12.22 of the County of Santa Cruz, California, Code requires various specified energy conservation measures in "multi-family buildings". This is defined to mean "any privately owned residential rental building, or portion thereof, which contains four units or more, and has a heating system; including apartments, flats, and all condominium conversions ..., but not including residential dwellings of less than four units, mobile homes or hotels, motels or inns whose primary purpose is to provide occupancy for less than 30 days".[22] The various specified energy conservation requirements refer to ceiling

---

circulating system must be insulated to an R-2. Shower heads must restrict flow to 3 gallons per minute. Air conditioner covers must be installed for the heating season.

8. All steam heating pipes not located in habitable rooms must be insulated to R-4. All hydronic heating pipes not located in habitable rooms must be insulated to R-2.
9. Floors over vented spaces or garages must be insulated.

18. Wis. Stat. Ann. § 101.122(2)(a).
19. *Id.*, § 101.122(4).
20. *Id.*, §§ 101.122(4)(b), (c), (6w).
21. Minn. Stat. Ann. c.116J.
22. Santa Cruz County Code § 12.22.030 (1985 Rev.).

insulation, weather-stripping, water-heater insulation blankets, low-flow showerheads, ducts and fitting joints, and caulking or sealing major cracks and joints in the building exterior.[23] The energy conservation requirements specified in the Code must be installed in connection with each transfer of title to every multi-family building unless one of the specified exemptions applies.[24] It is a violation of the Code for any person to sell or to offer to sell any multi-family building subject to the Code without full disclosure of the provisions of the Code, to refuse or neglect to install the basic energy conservation requirements within six months of the transfer of title of any multi-family dwelling, and to evict any tenant in retaliation for the exercise by the tenant of any rights created by the Code.[25]

The City of Portland, Oregon, passed an ordinance in 1979 requiring all rental housing to be retrofitted with "cost-effective" conservation measures by 1984.[26] "Cost-effective" was defined as meaning a payback period[27] of a maximum of ten years. The enforcement was to be by way of a deed restriction at the time of sale.

Existing Consumer Protection Legislation

One additional type of legislation that appears to promote energy efficiency in rental buildings is consumer protection legislation. The State of New York *Truth in Heating Act*,[28] which became effective on 1 January 1981, entitles all

---

23. *Id.*, § 12.22.070.
24. Miscellaneous exemptions are specified in § 12.22.050.
25. *Id.*, § 12.22.090(a).
26. Ordinance No.148,251 (15 August 1979). This ordinance is discussed *in* R.H. Counihan & D. Nemtzow, "Energy Conservation and the Rental Housing Market" (1981) 2 Solar Law Reporter 1103 at 1130; D. Talbot & R.E. Morgan, *Power and Light: Political Strategies for the Solar Transition* (New York: Pilgrim Press, 1981) at 177-191.
27. The "payback period" is the length of time it would take a purchaser of an energy conservation device to recoup the capital cost of the device by energy savings.
28. N.Y. Stat. § 17-103 (1983). The regulations adopted by the New York State Energy Office for implementing this Act are contained in 9 NYCRR § 7835.

prospective renters (and purchasers) of houses and residential apartments within the state to receive information on the past heating and cooling costs of a residence they are considering renting (or buying). The information which must be divulged to prospective renters/purchasers is: the name and address of the fuel vendor and/or utility providing heating/cooling fuel(s); the type, quantity and cost of all fuel and/or utility services consumed during the preceding two-year period for heating and cooling; any additional costs for lighting and water heating; whether other energy sources (such as solar, wind, or wood) were used in the residence; and, if the residence was unoccupied for thirty consecutive days or longer, the approximate dates during which it was unoccupied. A prospective tenant/purchaser may make a request for the relevant information either verbally or in writing, and the landlord must respond immediately. Landlords are required to request copies of the relevant bills from the utility and/or vendor within five days after being notified that a residence is to be vacated. Vendors then have ten days in which to respond to the landlord.[29]

The Preferred Policy for Rental Buildings

The policy issue to be determined is which party has the duty to install and maintain energy conservation measures sufficient for the premises to meet a designated standard of energy efficiency: tenants or landlords?

In the context of residential premises, the author submits that the legal duty to make rental buildings energy efficient should be the responsibility of the landlord. In light of the short duration of most residential leases,[30] it would seem unfair to expect the tenant to pay for the installation of energy conservation measures as the value of the reversion of the landlord in the property clearly outweighs the present interest of the tenant. A useful analogy here is the law relating to repairs of

---

29. Information supplied by Mr. William D. Cotter, Commissioner, New York State Energy Office.

30. See *supra*, note 4 and accompanying text. Note also that the vast majority of fixed-term leases for residential premises are for one year or less.

rented premises.[31] At common law, by virtue of the prevailing principle of *caveat emptor*,[32] in the absence of an express agreement to the contrary, there is no implied covenant that the premises are in good repair at the commencement of the tenancy or even that the premises are fit for human habitation.[33] In Canada and Australia, various reports have referred to the inequity of this law and have suggested that it is appropriate in light of the landlord's reversionary interest that the duty to put and maintain rented housing in a good state of repair should be imposed on the landlord.[34] This reform has been adopted in the majority of Canadian provinces and Australian states' residential tenancies statutes.[35] It is submitted that the arguments

---

31. *Williams and Rhodes Canadian Law of Landlord and Tenant*, 6th ed. (Toronto: Carswell, 1988) v.1, c.10.
32. As Erle CJ stated in *Robbins* v. *Jones* (1863), 15 C.B. (N.S.) 221 at 240; 143 E.R. 768 at 776, "Fraud apart, there is no law against letting a tumbledown house".
33. There are two exceptions to this rule. First, a warranty of fitness was implied where a tenancy agreement was entered into before the building of the premises was complete (*Miller* v. *Cannon Hill Estates Ltd.*, [1931] 2 K.B. 113). Second, there is an implied condition that in the case of furnished premises the premises are fit for human habitation at the commencement of the tenancy agreement (*Smith* v. *Marrable* (1843), 11 M & W 5; 152 E.R. 693; *Collins* v. *Hopkins*, [1923] 2 K.B. 617; *Pampris* v. *Thanos*, [1968] N.S.W.R. 56 (C.A.)). It is arguable that this second exception is limited by its nature to residential leases, and has been rendered redundant by the modern residential tenancies legislation. In Canada, however, it appears that the exception applies to commercial tenancies: *see Davey* v. *Christoff* (1916), 28 D.L.R. 447 (Ont. S.C., A.D.).
34. Australia: A.J. Bradbrook, *Poverty and the Residential Landlord-Tenant Relationship* (Canberra: Aust. Gov. Pub. Serv., 1975) at 20-21; R. Sackville, *Law and Poverty in Australia* (Canberra: Aust. Gov. Pub. Serv., 1975) at 63-64.

    Canada: Ontario. Law Reform Commission, *Interim Report on Landlord and Tenant Law Applicable to Residential Tenancies* (Toronto: 1968) c.5; British Columbia. Law Reform Commission, *Report on Landlord and Tenant Relationships: Residential Tenancies, Project No.12* (Victoria: 1973) c.7.
35. Residential tenancy statutes are a sub-species of landlord-tenant statutes.

    Canada: *Residential Tenancy Act*, S.B.C. 1984, c.15, s.8(1);

underlying this change to the law of repairs apply with equal force to the duty to ensure that rental premises are energy efficient.

In the context of commercial premises, the author submits that the legal duty to make the premises energy efficient should depend on the length of the tenancy. Where the tenancy is of short duration, it would seem reasonable to expect the landlord to pay for the installation of energy conservation measures because of the high value of the landlord's reversion in the property in comparison with the tenant's possessory interest. Conversely, where the tenancy is of long duration, it would seem reasonable to expect the tenant to pay the relevant costs. A useful precedent is the existing dividing or line fences legislation.[36] In many jurisdictions, this legislation authorizes the landlord to recover contributions towards the satisfaction of any liability incurred by them during the tenancy in respect of fencing work performed in relation to fences dividing the land occupied by the tenant from the land of adjoining owners.[37] In most cases, the Act specifies the actual contribution. For

---

Landlord and Tenant Act, R.S.M. 1987, c.L-70, s.95(1); *Residential Tenancies Act*, S.N.B. 1975, c.R-10.2, s.3; *Landlord and Tenant (Residential Tenancies) Act*, S.N. 1973, No.54, s.7(1); *Residential Tenancies Act*, S.N.S. 1970, c.13, s.6; *Landlord and Tenant Act*, R.S.O. 1980, c.232, s.96(1); *Landlord and Tenant Act*, R.S.P.E.I. 1974, c.L-7, s.102(1); *Residential Tenancies Act*, R.S.S. 1978, c.R-22, s.20(1). Note that the *Landlord and Tenant Act*, R.S.A. 1980, c.L-6, s.14, merely imposes an obligation on the landlord to ensure that the rented premises are habitable at the commencement of the tenancy. There is no continuing obligation on the landlord to maintain the premises in repair.

Australia: *Residential Tenancies Act 1987* (N.S.W.), s.25; *Residential Tenancies Act 1980* (Vic.), s.97; *Residential Tenancies Act 1975* (Qld.), s.7(a); *Residential Tenancies Act 1978* (S.A.), s.46; *Residential Tenancies Act 1987* (W.A.), s.42.

36. In Canada, for example, *see Line Fences Act*, R.S.A. 1980, c.L-16; *Line Fences Act*, R.S.O. 1980, c.L-22.

37. This is the format of the Australian legislation. *See Dividing Fences Act 1951* (N.S.W.), s.18; *Fences Act 1968* (Vic.), s.10; *Dividing Fences Act 1953* (Qld.), s.20; *Fences Act 1975* (S.A.), s.14; *Fences Act 1961* (W.A.), s.19; *Boundary Fences Act 1908* (Tas.), s.37; *Fences Act 1980* (N.T.), s.20.

example, the South Australian legislation states:

> The contribution recoverable under subsection (1) of this section shall be determined in accordance with the following principles:
>
> (a) where at the time of the performance of the fencing work a period of less than three years remains of the term of the tenancy, no contribution is payable by the tenant;
>
> (b) where at the end of the performance of the fencing work a period of at least three years but less than six years remains of the term of the tenancy, a contribution of one-quarter of the liability incurred by the landlord is payable by the tenant;
>
> (c) where at the time of the performance of the fencing work a period of at least six years but less than twelve years remains of the term of the tenancy, a contribution of one-half of the liability incurred by the landlord is payable by the tenant; and
>
> (d) where at the time of the performance of the fencing work a period of at least twelve years remains of the term of the tenancy, the landlord may recover a sum sufficient to satisfy his total liability from the tenant.[38]

The author recommends that the duty to make the premises energy efficient should be allocated on a similar basis.

What is the most appropriate means of implementing this recommended division of legal responsibility? The following sections analyze the existing legislation, and all other possible options for law reform, to determine the "preferred" (the most effective) legislation to encourage energy efficiency in rental buildings.

The Preferred Legislation

As can be seen, there is very little existing energy efficiency legislation specific to rental buildings. The existing legislation which appears to encourage rental building energy conservation is analyzed to determine its effectiveness, and some law reforms are suggested. The author notes that landlord-tenant legislation and general property legislation do not provide for energy efficiency in rental buildings, but recommends that they do so. The chapter concludes with a summary, which includes the author's proposed "preferred" legal management model to

---

38. *Fences Act 1975* (S.A.), s.14(2).

encourage energy efficiency in rental buildings.

Preferred Financial Incentives Legislation

The only existing financial incentives legislation relevant to energy conservation in rental buildings is income tax legislation. As tax relief for the cost of purchasing and installing energy conservation measures in rental buildings will usually not be available because of the wording of income tax legislation, it is submitted that the existing financial incentives legislation is not "preferred" legislation as it relates to energy conservation measures in rental buildings.

However, the author suggests that financial incentives could be used much more extensively to achieve energy efficiency in rental buildings. However, law reform will be required to establish new financial incentives through taxation.

Recommended Financial Incentives (Taxation) Reforms

The author recommends three types of financial incentives (taxation) reforms to encourage landlords to install energy conservation measures in their rental buildings: an income tax deduction or a credit/rebate for the cost of purchasing and installing energy conservation measures, an investment allowance for the installation of energy conservation measures, and municipal property tax[39] concessions.

Income Tax Deductions or Credit/Rebate

The first recommended taxation reform is an income tax deduction or a credit/rebate to landlords for the cost of purchasing and installing energy conservation measures in their rental buildings. It is submitted that landlords who invest in energy conservation measures in their rental buildings should be eligible for income tax relief in respect of the full cost of purchasing and installing designated energy conservation measures.

The concept of allowing tax relief under income tax

---

39. "Tax" in Canada and the United States; "rates" in United Kingdom and Australia.

legislation in respect of energy conservation measures is not novel. A useful precedent was s.82KM-82KS of the Australian *Income Tax Assessment Act 1936*, which provided in limited circumstances for a deduction for thermal insulation materials installed in private dwellings. Unfortunately, this provision was abolished by the federal Labour government shortly after it assumed office, and the incentive does not apply to payments made by a taxpayer after 23 August 1983.

The issue arises as to whether the tax relief should be couched as a deduction, a credit, or a rebate. This raises the same issues of tax equity as discussed in the context of financial incentives for the installation of energy conservation measures in new buildings.[40] As explained earlier, deductions would constitute the least equitable means of incentive, as the rich obtain greater value from the deduction than the poor.[41] A tax credit or rebate has the advantage that both constitute an absolute reduction in tax liability which does not vary in value according to the level of the taxpayer's income. This applies regardless of whether a full credit is given for the value of the energy conservation measure or whether the rebate is limited to the prevailing base marginal tax rate.

It is suggested that the cost of approved energy conservation measures installed by landlords in rental buildings should be classified as a full credit, entitling the landlord to one dollar tax relief for each dollar spent. Measures would not qualify for the proposed credit unless they were approved as meeting minimum building performance or efficiency standards.[42] Suitable standards would be developed from time to time by the relevant national standards association. Once

---

40. See *supra*, Chapter 2, notes 111-113 and accompanying text.
41. See *supra*, Chapter 2, note 113 and accompanying text. *See also* S.V. Hyatt, "Thermal Efficiency and Taxes: The Residential Energy Conservation Tax Credit" (1977) 14 Harvard J. Legis. 281 at 302-303.
42. Experience in the United States with the operation of various State and federal income tax incentives for renewable energy devices and energy conservation measures suggests strongly that performance or efficiency standards are essential. See, for example, Michigan Department of Commerce, *Michigan Tax Incentives for Residential Renewable Energy Systems*, December 1986.

published, the standards would be incorporated into law by means of the *Income Tax Regulations*. The onus would be placed on manufacturers to certify their energy conservation measures as complying with the Regulations so as to qualify for the credit.

The proposed tax rebate would be incorporated into the income tax legislation as a new section, and could be drafted as follows:

> (1) An amount paid by the taxpayer in the year of income for the purchase and installation of an energy conservation measure in any building owned by the taxpayer and leased in whole or in part to another person or persons for the purpose of producing assessable income shall, for the purposes of this section, be treated as a rebatable amount in respect of that year of income.
>
> (2) Where a rebatable amount is, or rebatable amounts are, applicable to a taxpayer in respect of a year of income, the taxpayer is entitled to a rebate of tax in the taxpayer's assessment in respect of income of that year of income of the full amount of the expenditure.

A new statutory definition of "energy conservation measure" would be required. This would preferably be included in the general definition section of the income tax legislation, and would read as follows:

> "energy conservation measure" means an installation or modification of an installation in a building which is primarily intended to reduce energy consumption, including:
>
> (a) insulation of the building structure and systems within the building;
> (b) storm windows and doors, multiglazed windows and doors, heat absorbing or heat reflective glazed and coated windows and door systems, additional glazing, reductions in glass area, and other window and door system modifications;
> (c) automatic energy control systems;
> (d) caulking and weather-stripping;
> (e) replacement or modification of lighting fixtures which increase the energy efficiency of the lighting system;
> (f) energy recovery systems; and
> (g) such other measures as, in the opinion of the Commissioner, will save a substantial amount of energy.[43]

---

43. This legislation is modelled on Wash. Rev. Code, s.43.19.670(2).

Investment Allowance

A second recommended taxation reform to encourage landlords to install energy conservation measures in their rental buildings is an investment allowance for the installation of energy conservation measures. This allowance could take the form of a 50 percent deduction of the costs of purchase and installation of energy conservation measures in the year of income in which the costs were incurred, with a further 50 percent deduction available in respect of the next succeeding year of income. The proposed allowance would be incorporated into the income tax legislation as a new section and could be drafted as follows:

> When a taxpayer has incurred expenditure of a capital nature on purchasing and installing an energy conservation measure in a building owned by the taxpayer and leased in whole or in part to another person or persons for the purpose of producing assessable income, there shall be allowed to the taxpayer as a deduction from the taxpayer's assessable income of the year of income during which the energy conservation measure is either used or installed ready for use, an amount equal to fifty (50) percentum of the expenditure incurred by the taxpayer, and a deduction of a further fifty (50) percentum of the expenditure incurred by the taxpayer shall be allowed to the taxpayer in respect of the next succeeding year of income.

Municipal Property Tax Concessions

A third recommended taxation reform to encourage landlords to install energy conservation measures in their rental buildings is to offer municipal property tax concessions. At present, there are two major systems of fixing municipal taxes: the site value system, and the improved capital value system. The majority of local councils now adopt the latter system. Unfortunately, the improved capital value system is counter-productive to the goal of encouraging the installation of energy conservation measures in rental buildings, as any such installation will have the effect of increasing the rates to the extent to which the installation increases the capital value of the property.

Two possible reforms should be considered. First, those councils which operate under the improved capital value system for taxes could disregard for rating purposes any increased value to the property caused solely by the installation of any energy conservation measures. In some jurisdictions this change could be effected by a decision of the local council, while in other jurisdictions changes to the local government legislation would be required. A second, and more radical, reform would be to offer a partial rebate from the taxes in respect of rented buildings in which designated energy conservation measures are installed. Precedents exist for rebates for particular types of property. Perhaps the most wide-embracing rebate is the 45 percent rebate offered in Australia by the City of Adelaide to all owner-occupiers of residential property throughout the city.

Although specific property tax incentives have not yet been employed anywhere in Canada or Australia, they are widespread in the United States.[44] The origins of the incentives was a study undertaken in 1974-1977 by the American Bar Foundation, which recommended that states should exempt renewable energy devices and energy conservation measures from property taxes on the basis that this would encourage investment in such devices and measures without imposing any additional financial burdens on local communities.[45] Some U.S. states have adopted this option and apply a real property tax exemption for the cost of renewable energy devices or energy conservation measures as an alternative to an income tax credit or deduction.[46]

---

44. For a discussion of the position in the United States, *see* H. Lee, "The Role of Local Governments in Promoting Energy Efficiency" (1981) 6 Ann Rev. Energy 309 at 313*ff.*
45. W. Thomas, A. Miller & R. Robbins, *Overcoming Legal Uncertainties About Use of Solar Energy Systems* (American Bar Foundation, 1978). *See also* Environmental Law Institute, *Legal Barriers to Solar Heating and Cooling of Buildings* (Washington, D.C.: Books for Business, Inc., 1981) at 260-269; S.E. Ferrey, "Solar Banking: Constructing New Solutions to the Urban Energy Crisis" (1981) 18 Harvard J. Legis. 483 at 511-512.
46. For example, s.487 of the New York State Real Property Tax provides a 15-year real property tax exemption for eligible solar and wind devices constructed prior to 1 July 1988. *See also* Ill Rev Stat, s.50ld-2.

Preferred Energy Conservation Building Standards Legislation for Rental Buildings

Should special laws for energy conservation measures in rental buildings be enacted in building standards legislation? For various reasons, it is submitted that this is not a desirable approach. The creation of new building standards for rental buildings along the lines of the U.S. legislation referred to earlier would involve the creation of a new bureaucracy to administer the new laws, which is a costly and inefficient method of tackling the problem. In addition, it might be queried why rental premises should be singled out for special attention in the building regulations. If there is a lack of energy conservation features in buildings, surely this problem should be addressed for all buildings in the jurisdiction, regardless of the nature of their occupancy. For similar reasons it is submitted that the enactment of special laws regarding energy conservation measures in rental buildings in local government ordinances is also undesirable. Municipalities are not financially or administratively equipped to conduct the additional compliance work which would be required, and it would seem socially undesirable that different standards should apply in different areas of the jurisdiction and in different municipalities in large cities. Builders and developers could validly complain of the inevitable confusion that would be caused by this system. In light of these various arguments, it would seem that the existing building standards legislation is not "preferred" legislation as it relates to energy conservation in rental buildings, and that law reform is not desirable.

Preferred Consumer Protection Legislation

This leaves the question whether consumer protection legislation similar to the New York *Truth in Heating Act* should be enacted by other jurisdictions. It is interesting to note that similar legislation already exists in many countries in the context of legislation controlling the sale of land and second-hand motor vehicles.[47] This legislation requires vendors to supply

---

47. See, for example, Australia: *Sale of Land Act 1962* (Vic.), s.32; *Land and Business Agents Act 1973* (S.A.), s.90; *SecondHand Motor Vehicles Act 1983* (S.A.), s.18. *See also* J. Goldring, L. Maher & J. McKeough,

prospective purchasers with specified information designed to assist purchasers in making their decision whether or not to proceed with the purchase. It can be argued that consumer protection legislation like the *Truth in Heating Act* is merely a logical extension of legislation which has proven successful in other social contexts. Despite this argument, although the *Truth in Heating Act* would certainly be effective in informing tenants of their likely financial obligations in respect of energy use, such legislation appears to be inadequate for two major reasons. First, the information received from landlords may be very misleading. The circumstances of the earlier tenancy may have been quite different. For example, the premises may have been vacant for long periods during the previous tenancy, the previous tenant's family may have used thermostat settings that were unusually high or low, the past winters or summers covered by the data may have been unusually severe or mild, or energy conservation improvements may have been made to the residence since the last heating or cooling season. Second, legislation such as the *Truth in Heating Act* does not require any energy conservation improvements to be made. At best, all it does is to put indirect pressure on landlords to upgrade energy conservation measures in their premises. If vacancy rates for rented accommodation are low, any such pressure is nullified. Any new legislation should surely tackle the problem of improving the energy efficiency of rental buildings more directly. In light of these various arguments, it would seem that the existing United States' consumer protection legislation is not "preferred" as it relates to energy conservation in rental buildings.

Ironically, all the legislation that *appears* to encourage energy conservation in rental buildings – financial incentives, building standards, and consumer protection – does not do so. Energy conservation is not mentioned in the legislation which governs residential or commercial landlord-tenant relationships, but the author recommends amending these statutes, so that they would become "preferred legislation".

Consumer Protection Law in Australia, 3d ed. (Sydney: Butterworths, 1987) para.5109.

Preferred Landlord-tenant and General
Property Legislation

One might expect that the existing legislation which governs residential or commercial landlord-tenant relationships would encourage the installation of energy conservation. However, energy conservation is not mentioned by the provincial/state legislation which governs Canadian and Australian *residential* tenancies or by the common law of landlord and tenant. It must be noted that while "landlord-tenant" legislation governs *residential* landlord-tenant relationships, no equivalent legislation governs the *commercial* landlord-tenant relationship. However, as general property statutes do contain some tenancy provisions, the author suggests that general property statutes be amended to include energy conservation law reforms for commercial rental buildings. The author submits that landlord-tenant and general property legislation, which govern residential and commercial landlord-tenant relationships, respectively, be amended to encourage energy efficiency.

Recommended Landlord-tenant Law Reforms: Residential Premises

In recent years, all Canadian provinces and Australian states (except Tasmania) have codified the law relating to residential tenancies.[48] Any reforms designed to promote the use of energy

---

48. Residential tenancy statutes are a sub-species of landlord-tenant statutes. The Canadian and Australian residential tenancies statutes are:

Canada: *Landlord and Tenant Act*, R.S.O. 1980, c.232; *Residential Rent Regulation Act*, S.O. 1986, c.63; *Landlord and Tenant Act*, R.S.A. 1980, c.L-6; *Residential Tenancy Act*, S.B.C. 1984, c.15; *Landlord and Tenant Act*, R.S.M. 1987, c.L-70; *Residential Rent Regulation Act*, R.S.M. 1987, c.R-118; *Residential Tenancies Act*, S.N.B. 1975, c.R-10.2; *Residential Tenancies Act*, S.N. 1988, c.44; *Residential Tenancies Act*, S.N.S. 1970, c.13; *Rent Review Act*, S.N.S. 1975, c.56; *Rental of Residential Property Act*, S.P.E.I. 1988, c.58; *Residential Tenancies Act*, R.S.S. 1978, c.R-22; *Act Respecting the Régie du logement*, R.S.Q., c.R-8.1.

Australia: *Residential Tenancies Act 1987* (N.S.W.); *Residential Tenancies Act 1980* (Vic.); *Residential Tenancies Act 1975* (Qld.); *Residential Tenancies Act 1978* (S.A.); *Residential Tenancies Act 1987*

conservation measures in rental housing would most appropriately be incorporated as additional sections in each province's or state's Act.

It is submitted that three residential landlord-tenant law reform measures be considered: that rent control legislation take energy conservation into account, that a statutory duty be imposed on landlords to comply with minimum prescribed energy efficiency standards, and that tenants be reimbursed by landlords for the cost of energy conservation measures added by them during the tenancy agreement.

The first recommended residential landlord-tenant law reform is that energy conservation be taken into account in the rent control legislation. The residential tenancies statutes in most jurisdictions have established a new system of controlling rents. The Canadian provinces where rent control legislation exists[49] have limited their statutory scheme to a review of rent increases in excess of the annual percentage increase allowed by regulation. Although the legislation varies between the provinces, the Nova Scotia *Rent Review Act*[50] is a typical illustration. Pursuant to s.9, a landlord may increase the rent of residential premises in excess of the authorized amount after having given three months' notice to the tenant, upon application to the residential tenancy officer at least two months prior to the date of the rent increase, and upon being authorized by an order of the residential tenancy officer. Section 10(2) states that in determining the issue of a rent increase, the residential tenancy officer must consider:

> (a) any increase in rent for the residential premises which took effect on or before the thirty-first day of December, 1975;
> (b) return on investment;
> (c) whether or not the increase in rent sought by the landlord is necessary in order to prevent the landlord from being adversely affected or sustaining a financial loss in the operation of the building in which the residential premises are situate; and
> (d) such other matters as may be prescribed by the regulations.

(W.A.).

49. There is no rent control legislation in Alberta.

50. *Rent Review Act*, R.S.N.S. 1989, c.398.

It is submitted that this subsection and its equivalent in the other provinces' legislation should be amended so as to include the provision of energy conservation measures in the rental premises as a relevant factor in determining whether the rent is excessive. This provision, which would become the new s.10(2)(d)[51], could be drafted as follows: "(d) any energy conservation measures provided by the landlord in the rented premises." If this reform were adopted, a new statutory definition of "energy conservation measure" would be required. This would be included in the general definition section of the legislation, and would be drafted in the same manner as the definition suggested earlier for inclusion in the income tax legislation.[52]

Australian landlord-tenant provisions are similar to, but not identical to, those of Canada. In Australia, the statutory scheme for controlling rents is more wide-ranging.[53] For example, in Victoria, s.63(1) of the *Residential Tenancies Act 1980* states that a tenant who considers that the present rent is excessive having regard to the fact that the landlord has reduced or withdrawn any goods, services or facilities provided with the rented premises, or who has received notice from the landlord of a rent increase and considers that the increase is excessive, may request the state Director of Consumer Affairs to investigate and report on the matter. Once the Director of Consumer Affairs has reported on the matter, s.64(1) permits the tenant to apply to the Residential Tenancies Tribunal for an order declaring the rent or proposed rent to be excessive.

Section 63(2) specifies the relevant considerations for the Tribunal to take into account when a tenant makes an application under s.63(1):

> 63(2) For the purposes of this section, rent under a tenancy agreement shall be regarded as excessive if it is significantly more than the rent payable for comparable rented premises let under a tenancy agreement by a landlord, other than a public statutory

---

51. The existing "s.10(2)(d)" would become "s.10(2)(e)".
52. See *supra*, note 43 and accompanying text.
53. For a discussion of the Australian legislation, *see* A.J. Bradbrook, S.V. MacCallum & A.P. Moore, *Residential Tenancy Law and Practice* (Sydney: Law Book Co., 1983) paras.1209-1228.

authority, in the locality of the first-mentioned rented premises and the following factors shall also be taken into account:

(a) The cost of goods, services and facilities provided with the rented premises;
(b) Any charges in respect of the rented premises for which the landlord is or may be liable by or under this or any other Act or the tenancy agreement;
(c) The state of repair and general condition of the rented premises;
(d) Any goods, services or facilities provided by the tenant under the tenancy agreement or any charges payable by the tenant under this or any other Act or the tenancy agreement;
(e) Any work which the tenant has, with the landlord's consent, done or has agreed with the landlord to do in relation to the premises; and
(f) Any valuation of the rented premises.[54]

It is submitted that this subsection should be amended so as to include the provision of energy conservation measures in the rental premises as a relevant factor in determining whether the rent is excessive. The provision, which would become s.63(2)(g), could be drafted similarly to the amendment proposed in respect of the Canadian provinces.[55]

The second recommended residential landlord-tenant law reform, a far more radical, but effective, means of increasing the use of energy conservation measures in rental housing, would be to impose a statutory duty on landlords to comply with minimum prescribed standards of energy efficiency. Residential tenancies legislation already imposes a number of statutory duties on landlords covering a wide range of issues, such as repairs, cleanliness, quiet enjoyment, rights of entry, and ensuring that the tenant obtains vacant possession at the commencement of the lease. This proposal would involve the creation of an additional statutory duty relating to energy efficiency.

Under this proposal, minimum energy performance or efficiency standards will need to be prescribed. Suitable standards would be developed from time to time by the relevant national standards association. Once published, the standards

---

54. There is no comparable legislation in Queensland.
55. See the text at note 42, *supra*.

would be incorporated into law by means of the residential tenancies regulations.

The legislation could be drafted as follows: "A landlord under a tenancy agreements shall ensure that the rented premises comply with minimum prescribed energy efficiency standards contained in the Regulations." The regulation-making power in the residential tenancies legislation would also need to be amended to allow for the incorporation of the energy efficiency standards into law.

In the event a landlord breaches the new statutory duty to ensure that rental premises comply with minimum prescribed energy efficiency standards contained in the residential tenancies regulations, it is submitted, as a third residential landlord-tenant law reform recommendation, that the tenant should be permitted to arrange for the installation of the necessary energy conservation measures and to seek reimbursement from the landlord. Such reimbursement should be available as a last resort by order of the Residential Tenancies Board/Tribunal.

At first glance, it may seem unfair that a tenant should legally be able to install energy conservation measures in the rental premises and then to demand reimbursement by the landlord. However, on closer analysis, there are three possible justifications for law reform of this nature.

The first justification is that a similar remedy is applied under the residential tenancies legislation in some jurisdictions where the landlord breaches his duty to maintain the premises in a good state of repair. For example, pursuant to s.99 of the *Residential Tenancies Act 1980* (Vic.), where a tenant is unable, after taking reasonable steps, to make arrangements for the carrying out forthwith by the landlord or the landlord's agent of urgent repairs to the rental premises, the tenant may carry out the repairs, and the landlord is liable to reimburse the tenant for the reasonable costs of the repairs or $500, whichever is the less.[56] This is a useful and close analogy to the issue under discussion.

---

56. *C.f.*, *Residential Tenancies Act 1987* (N.S.W.), s.28 ($800 maximum limit); *Residential Tenancies Act 1987* (W.A.), s.43 (no maximum specified).

Secondly, although at common law a tenant is not permitted to remove agricultural fixtures installed during the term of the tenancy,[57] agricultural holdings legislation in some jurisdictions allows the tenant a limited right to claim compensation at the termination of the tenancy from the landlord for certain specified types of improvement to the extent to which the improvement fairly represents the value of the improvement to an incoming tenant.[58] This is a precedent for the proposition that in certain circumstances a landlord should be legally required to pay the tenant for the value of improvements made by the tenant during the tenancy. The common law rule exempting the landlord from such a duty in all cases should no longer be considered to be equitable or appropriate.

Thirdly, there are strong public policy arguments in favour of adopting all reasonable legal avenues to maximize the use of energy conservation measures. As discussed earlier, the major argument is that the use of such measures can assist in the conservation of fossil fuels and reduce pollution levels resulting from conventional sources of electricity production. Additional relevant factors are global warming, the need to minimize the balance of trade deficit by importing less oil, and the reduction in the cost of living which the use of energy conservation measures would entail.[59]

To be equitable to the landlord, it is submitted that any provision requiring the tenants to be reimbursed for the cost of energy conservation measures added by them during the tenancy agreement should specify a maximum figure for the liability of the landlord to reimburse the tenant, should not apply unless the tenant has given the landlord a reasonable opportunity to arrange for the installation of specified measures, and should specify the types of measures which are subject to the operation of the legislation.

---

57. *Elwes* v. *Maw* (1802), 3 East 38; 102 E.R. 510.

58. See, e.g., *Agricultural Holdings Act 1941* (N.S.W.). ss.7-12; *Property Law Act 1974* (Qld.), ss.153-167; *Agricultural Holdings Act 1891* (S.A.), ss.6-22.

59. *See* A.J. Bradbrook, *Solar Energy and the Law* (Sydney: Law Book Co., 1984) para.805.

The provision could be drafted as follows:

(1) Where the tenant under a tenancy agreement is unable, after taking reasonable steps, to make arrangements for the purchase and installation by the landlord or an agent of the landlord within a reasonable time of any of the prescribed energy conservation measures in the rented premises —

(a) the tenant may install the energy conservation measures; and

(b) subject to subsection (2), the landlord is liable to reimburse the tenant for the reasonable costs of the purchase and installation of the energy conservation measures (up to but not exceeding, in each case, $1,000 or such other amount as may be prescribed) incurred by the tenant in purchasing and installing energy conservation measures.

(2) A landlord is not liable to reimburse an amount to a tenant under subsection (1) until the expiration of a period of fourteen days after the tenant gives the landlord notice in writing specifying the energy conservation measures purchased and installed and the costs thereby incurred.

The remedy for the tenant, if the landlord fails to comply with his statutory duty to reimburse the tenant for the cost of purchasing and installing energy conservation measures, would be to lodge a claim for compensation with the Residential Tenancies Board/Tribunal. This would require no legislative change, as the existing provisions relating to compensation claims are sufficiently broad to encompass the issue under discussion.

Recommended Property Law Reforms: Commercial Premises

Unlike the law of residential tenancies, there is no legislation which applies to all commercial premises onto which specific statutory requirements relating to energy efficiency laws can be grafted. It is suggested that any energy conservation law reforms would most appropriately be inserted as a new section in the general property statutes in force in each jurisdiction. The general property legislation already contains miscellaneous provisions relating to tenancies, so the addition of energy conservation provisions would not appear incongruous or inappropriate.

It is submitted that two amendments be made to general property statutes regarding the commercial landlord-tenant relationship: that a statutory duty be imposed on landlords to comply with minimum prescribed energy efficiency standards, and that commercial tenants be reimbursed by landlords for the cost of energy conservation measures added by them during the tenancy agreement.

First, for reasons discussed earlier regarding residential tenancies,[60] it is submitted that the general property law should impose a statutory duty on commercial landlords to comply with minimum prescribed energy efficiency standards, but should also impose a duty of contribution on the commercial tenant which will vary according to the length of the tenancy. The proposed legislation would read as follows:

(1) A landlord under a commercial tenancy agreement shall ensure that the rented premises comply with minimum prescribed energy efficiency standards contained in the Regulations.

(2) Subject to any agreement between a landlord and tenant and the provisions of this section, the landlord may recover, as debt due to him or her, contribution towards the satisfaction of any liability incurred by the landlord during the tenancy in complying with minimum prescribed energy efficiency standards.

(3) The contribution recoverable under subsection (2) of this section shall be determined in accordance with the following principles:

(a) where at the time of complying with minimum energy efficiency standards a period of less than three years remains of the term of the tenancy, no contribution is payable by the tenant;

(b) where at the time of complying with minimum energy efficiency standards a period of at least three years but less than six years remains of the term of the tenancy, a contribution of one-quarter of the liability incurred by the landlord is payable by the tenant;

(c) where at the time of complying with minimum energy efficiency standards a period of at least six years but less than twelve years remains of the term of the tenancy, a contribution of one-half of the liability incurred by the landlord is payable by the tenant; and

(d) where at the time of complying with minimum energy

60. See the "second residential landlord-tenant law reform recommendation" in the section of this chapter entitled "Recommended Landlord-tenant Law Reforms: Residential Premises".

> efficiency standards a period of at least twelve years remains of the term of the tenancy, the landlord may recover a sum sufficient to satisfy his total liability from the tenant.

As in the case of residential tenancies, minimum commercial tenancy energy efficiency standards will need to be prescribed. These standards would be developed by the relevant national standards association and would be incorporated into law by means of regulations made under the general property legislation. The legislation would read:

> The Governor in Council may make regulations for or with respect to prescribing minimum energy performance or energy efficiency standards for rented premises.

Second, it is further recommended that if a landlord breaches the statutory duty to ensure that the rented premises comply with the minimum prescribed energy standards, the commercial tenant should be permitted to arrange for the installation of the necessary energy conservation measures and to seek reimbursement from the landlord. The legislation required to effect this change would be drafted similarly to the provision recommended earlier in respect of residential premises.[61]

Summary

As has been shown, none of the existing financial incentives, building standards, or consumer protection legislation does directly encourage energy conservation in rental buildings. The author, therefore, submits that the preferred model to encourage energy efficiency in rental buildings would combine financial incentives (in the form of taxation measures for landlords) and legislation governing the residential and commercial landlord-tenant relationship. This is an application of the "carrot and stick" approach: the taxation measures will provide the incentive, while the tenancy law reforms will provide the necessary legal duty and enforcement mechanisms.

---

61. See the "third residential landlord-tenant law reform recommendation" in the section of this chapter entitled "Recommended Landlord-tenant Law Reforms: Residential Premises".

4. A SUMMARY OF RECOMMENDED LEGISLATIVE CHANGES

This chapter provides a brief overview of the existing energy conservation legislation for new, existing, and rental buildings, and highlights the recommended legislative changes.

Existing Energy Conservation Legislation for New and Existing Buildings

1. The existing legislation to encourage energy conservation in new and existing buildings includes:
   - financial incentives (stimulation) and
   - building standards (regulation).

2. The financial incentives legislation includes:
   - tax concessions,
   - government-backed or government grants or loans to individuals and businesses, and
   - statutory programs requiring energy utilities to improve energy efficiency in the community;
   - in addition, specific funded programs apply to existing buildings.

   It is submitted that tax incentives are the preferred option, however, law reform will be required to establish new financial incentives through taxation.

   IT IS RECOMMENDED THAT, if tax incentives are adopted, the following four policy issues should be addressed:

   i. ***that relevant criteria for the formulation of the income tax incentives be determined***

      Criteria ought to include the requirements that:
      - the tax incentives must be effective for the time period necessary for commercialization;

- the incentives must be adequate in level to stimulate commercial investment in the most viable energy conservation technology options;
- the incentives must provide for smooth, predictable phasing-out;
- the incentives must be selective for high performance and long operating life;
- the incentives must stimulate, rather than inhibit, technical progress and innovation;
- the potential for abuse must be minimized;
- the incentives must be simple to implement and administer;
- the incentives must address the financial imbalance in incentives between conventional technology and energy conservation technology; and
- the incentives must be market-oriented rather than technology-oriented.

ii. ***that building performance criteria be established as a condition precedent to any income tax incentives,***

iii. ***that new taxation incentives should constitute a credit or rebate, rather than a deduction under the terms of the income tax legislation,***

iv. ***and that any new taxation incentives should not remain in existence indefinitely, but should only be maintained until energy conservation measures are able to compete in the market place on an equal basis with conventional energy technologies.***

3. The five types of existing energy conservation building standards for new buildings are:
   - mandatory building performance regulations,
   - mandatory prescriptive regulations for building materials,

- mandatory or voluntary energy rating or labelling systems,
- a mandatory system of assigning energy and resource conservation option points to buildings,
- or a mandatory system requiring the vendor of the building to provide information in the contract of sale.

IT IS RECOMMENDED THAT the preferred model for energy conservation regulations in respect of new buildings should consist of a combination of building performance-based regulations and a conservation options points system.

4. The application of energy conservation regulations to existing buildings is problematic because it is often difficult, if not impossible, to add such measures into buildings once they have been erected. A further difficulty is that it may be regarded as politically unacceptable to apply retrospectively new regulations to existing buildings.

IT IS RECOMMENDED THAT reliance must primarily be based on financial incentives to encourage the installation of energy conservation measures in existing buildings on a voluntary basis.

IT IS FURTHER RECOMMENDED THAT this system be coupled with a system of compulsory disclosure of existing conservation measures whenever a building is sold or leased.

Existing Energy Conservation Legislation for Rental Buildings

5. The background policy issue that sets rental buildings apart from new and existing buildings as it relates to energy conservation measures is this: which party (landlord or tenant) has the duty to install and maintain energy conservation measures sufficient for the rental premises to

meet a designated standard of energy efficiency? Law reform will be required to establish the division of legal responsibility.

> ***IT IS RECOMMENDED THAT legislation make the energy efficiency of residential premises the legal responsibility of the landlord.***

> ***IT IS FURTHER RECOMMENDED THAT the legal responsibility to make commercial premises energy efficient should depend on the length of the tenancy:***
> - where the tenancy is of short duration, it would seem reasonable to expect the landlord to pay for the installation of energy conservation measures, because of the high value of the landlord's reversion in the property in comparison with the tenant's possessory interest.
> - conversely, where the tenancy is of long duration, it would seem reasonable to expect the tenant to pay the relevant costs.

6. Although existing financial incentives (stimulation), building standards (regulation), and consumer protection legislation *appear* to encourage energy conservation in rental buildings, they do not do so. Law reform will be required to compel landlords, residential tenants, and commercial tenants to install energy conservation measures into rental buildings.

> ***IT IS RECOMMENDED THAT statutory income tax incentives and landlord-tenant legislation impose on landlords the responsibility to install energy conservation measures into rental buildings.***

> ***Income tax incentives for landlords:***
> There are 3 possible income tax incentives which would encourage *landlords* to install energy conservation measures into rental buildings:
> - income tax deductions or credit/rebates,

- investment allowances,
- or municipal property tax concessions.

First, an income tax deduction or a credit/rebate should be allowed to the landlord for the cost of purchasing and installing energy conservation measures:

- for reasons of tax equity, it is suggested that the cost of approved energy conservation measures installed by landlords in rental buildings be classified as a full rebate, entitling the landlord to dollar tax relief for each dollar spent.
- measures would not qualify for the proposed rebate unless they were approved as meeting minimum performance or efficiency standards.
- suitable standards would be developed from time to time by the relevant standards association.
- once published, the standards ought to be incorporated into law by means of income tax regulations.
- the onus would be placed on manufacturers to certify their energy conservation measures as complying with the regulations so as to qualify for the rebate.

As a second possible alternative to the creation of an income tax rebate to landlords for the cost of purchasing and installing energy conservation measures in their rental premises, consideration should be given to the establishment of a special incentive in the form of an investment allowance:

- this allowance could take the form of a 50 percent deduction of the costs of purchase and installation of energy conservation measures in the year of income in which the costs were incurred, with a further 50 percent deduction available in respect to the next succeeding year of income.

As a third possible alternative to the creation of an income tax rebate to landlords for the cost of purchasing and installing energy conservation measures in their rental premises, consideration should be given to the possibility of municipal rates concessions.

Landlord-tenant legislation:
Landlord-tenant legislation could encourage *landlords* to install energy conservation measures into rental buildings:

- conservation measures should be taken into account in the excessive rent legislation, and
- a new statutory duty should be imposed on landlords to comply with minimum prescribed energy efficiency standards.

IT IS RECOMMENDED THAT <u>residential</u> <u>tenants</u> be encouraged to install energy conservation measures into rental buildings by means of landlord-tenant legislation:

- residential tenants should be reimbursed for the cost of energy conservation measures added by them during the tenancy agreement.
- these rights of reimbursement should specify:
  - a maximum figure for the liability of the landlord to reimburse the tenant,
  - should not apply unless the tenant has given the landlord a reasonable opportunity to arrange for the installation of specified measures,
  - and should specify the types of measures which are subject to the operation of the legislation.
- the remedy for the residential tenant, if the landlord fails to comply with the new statutory duty to ensure that the rented premises comply with minimum prescribed energy standards, would be to lodge a claim for compensation with the Residential Tenancies Tribunal.

IT IS RECOMMENDED THAT commercial tenants be encouraged to install energy conservation measures into rental buildings by means of general property statues:

- as there is no landlord-tenant legislation which applies to all commercial premises onto which specified requirements relating to energy efficiency laws can be grafted, it is suggested that new law reforms in this area would most appropriately be inserted as a new section in the general property statutes in force in each jurisdiction.
- it is suggested that imposing a statutory duty on commercial landlords to comply with minimum prescribed energy efficiency standards is not the most appropriate type of law reform, but to impose a duty of contribution on the tenant which will vary according to the length of the tenancy. As in the case of residential tenancies:
  - new minimum energy performance or efficiency standards will need to be prescribed.
  - these standards would be developed by the relevant standards association and would be incorporated into law by means of regulations made under the general property legislation.
  - if a landlord breaches the new statutory duty to ensure that the rented premises comply with minimum prescribed energy standards, the commercial tenant should be permitted to arrange for the installation of the necessary energy conservation measures and to seek reimbursement from the landlord.

INSTITUTE PUBLICATIONS

Following is a complete list of CIRL publications currently available:

Energy Conservation Legislation for Building Design and Construction, by Adrian J. Bradbrook. 1992. ISBN 0-919269-38-9. 88 p. $20.00

Resource Development and Aboriginal Land Rights, by Richard H. Bartlett. 1991. ISBN 0-919269-33-8. 122 p. $25.00

Managing Interjurisdictional Waters in Canada: A Constitutional Analysis, by Steven Alexander Kennett. 1991. ISBN 0-919269-31-1. 238 p. $26.00

Security of Title in Canadian Water Rights, by Alastair R. Lucas. 1990. ISBN 0-919269-22-2. 102 p. $22.00

The Offshore Petroleum Regimes of Canada and Australia, by Constance D. Hunt. 1989. ISBN 0-919269-29-X. 169 p. $14.40

The Inuvialuit Final Agreement, by Janet M. Keeping. 1989. ISBN 0-919269-28-1. 160 p. $14.40

Toxic Water Pollution in Canada: Regulatory Principles for Reduction and Elimination with Emphasis on Canadian Federal and Ontario Law, by Paul Muldoon and Marcia Valiante. 1989. ISBN 0-919269-26-5. 120 p. $22.00

Interjurisdictional Issues in Canadian Water Management, by J. Owen Saunders. 1988. ISBN 0-919269-27-3. 130 p. $22.00

The Framework of Water Rights Legislation in Canada, by David R. Percy. 1988. ISBN 0-919269-21-4. 103 p. $20.00

Maritime Boundaries and Resource Development: Options for the Beaufort Sea, by Donald R. Rothwell. 1988. ISBN 0-919269-24-9. 61 p. $9.00

A Reference Guide to Mining Legislation in Canada (Second Edition), by Barry Barton, Barbara Roulston and Nancy Strantz. 1988. ISBN 0-919269-25-7. 123 p. $18.00

Liability for Drilling- and Production-source Oil Pollution in the Canadian Offshore, by Christian G. Yoder. Working Paper 12. 1986. ISBN 0-919269-20-6. 84 p. $10.20

A Guide to Appearing Before the Surface Rights Board of Alberta (Second Edition), by Barry Barton and Barbara Roulston. Working Paper 11. 1986. ISBN 0-919269-19-2. 124 p. $10.20

Crown Timber Rights in Alberta, by N.D. Bankes. Working Paper 10. 1986. ISBN 0-919269-17-6. 128 p. $10.20

The Canadian Regulation of Offshore Installations, by Christian G. Yoder. Working Paper 9. 1985. ISBN 0-919269-18-4. 116 p. $10.20

The Assignment and Registration of Crown Mineral Interests, by N.D. Bankes. Working Paper 5. 1985. ISBN 0-919269-11-7. 126 p. $10.20

Oil and Gas Conservation on Canada Lands, by Owen L. Anderson. Working Paper 7. 1985. ISBN 0-919269-16-8. 122 p. $10.20

Canadian Electricity Exports: Legal and Regulatory Issues, by Alastair R. Lucas and J. Owen Saunders. Working Paper 3. 1983. ISBN 0-919269-09-5. 40 p. $5.70

The International Legal Context of Petroleum Operations in Canadian Arctic Waters, by Ian Townsend Gault. Canadian Continental Shelf Law 2; Working Paper 4. 1983. ISBN 0-919269-10-9. 76 p. $5.40

Acid Precipitation in North America: The Case for Transboundary Cooperation, by Douglas M. Johnston and Peter Finkle. 1983. ISBN 0-919269-05-2. 75 p. $6.00

Resources: The Newsletter of the Canadian Institute of Resources Law. ISSN 0714-5918. Quarterly. Free

Annual Report. Free

ESSAYS FROM THE INSTITUTE'S CONFERENCE ON NATURAL RESOURCES LAW

Growing Demands on a Shrinking Heritage: Managing Resource-use Conflicts, Essays from the Fifth Institute Conference on Natural Resources Law, Ottawa, Ontario, May 9-11, 1991. Monique Ross and J. Owen Saunders, eds. ISBN 0-919269-35-4. 431 p. (hardcover) $80.00

The Legal Challenge of Sustainable Development, Essays from the Fourth Institute Conference on Natural Resources Law, Ottawa, Ontario, May 10-12, 1989. J. Owen Saunders, ed. ISBN 0-919269-32-X. 401 p. (hardcover) $75.00

Trading Canada's Natural Resources, Essays from the Third Institute Conference on Natural Resources Law, Banff, Alberta, May 6-9, 1987. J. Owen Saunders, ed. 367 p. (hardcover) $75.00*

Managing Natural Resources in a Federal State, Essays from the Second Institute Conference on Natural Resources Law, Banff, Alberta, April 17-20, 1985. J. Owen Saunders, ed. 372 p. (hardcover) $70.00*

Public Disposition of Natural Resources, Essays from the First Institute Conference on Natural Resources Law, April 12-15, 1983. Nigel Bankes and J. Owen Saunders, eds. ISBN 0-919269-14-1. 366 p. (hardcover) $28.20

*20% off if you order two volumes of Essays from the Institute Conference on Natural Resources Law from Carswell. See "Other Publications" section for ordering information.

DISCUSSION PAPERS

Successor Liability for Environmental Damage, by Terry R. Davis. 1989. 46 p. $10.00

Surrounding Circumstances and Custom: Extrinsic Evidence in the Interpretation of Oil and Gas Industry Agreements in Alberta, by David E. Hardy. 1989. 38 p. $10.00

Views on Surface Rights in Alberta, Papers and materials from the Workshop on Surface Rights, Drumheller, 20-21 April 1988. Barry Barton, ed. 77 p. $10.00

Classifying Non-operating Interests in Oil and Gas, by Eugene Kuntz; Presented at a seminar sponsored by the Faculty of Law and the Canadian Institute of Resources Law, The University of Calgary, 7 April 1988. 31 p. $10.00

Publications are available from: Canadian Institute of Resources Law, 430 BioSciences Building, Faculty of Law, The University of Calgary, Calgary, Alberta, Canada T2N 1N4. Telephone (403) 220-3200. Facsimile (403) 282-6182. Telex 03-821545.

Orders from within Canada please add 7% GST. Orders from outside Canada please add $2.00 per book for postage and handling.

OTHER PUBLICATIONS

Canada Energy Law Services.
Canada Energy Law Service: Federal, 2 vols.
Canada Energy Law Service: Alberta, 1 vol.
ISBN 0-88820-108-7.
Available from: Carswell
Thomson Professional Publishing
Corporate Plaza
2075 Kennedy Road
Scarborough, Ontario, Canada
M1T 3V4

For more information call:
Toronto: (416) 609-8000
Ontario and Quebec: 1-800-387-0142
Other provinces: 1-800-387-5164
Fax: (416) 298-5094